Data Ethics of Power

Dedicated to Clara, Francesco, Tina, Bo and my dear friend Gry who showed me what ethics is in practice in each of their uniquely human ways.

Dedicato a Clara, Francesco, Tina, Bo e Gry che mi hanno mostrato ognuno a loro modo cosa sia l'etica nella vita quatodiana.

Data Ethics of Power

A Human Approach in the Big Data and AI Era

Gry Hasselbalch

Edward Elgar
PUBLISHING

Cheltenham, UK • Northampton, MA, USA

Published by
Edward Elgar Publishing Limited
The Lypiatts
15 Lansdown Road
Cheltenham
Glos GL50 2JA
UK

Edward Elgar Publishing, Inc.
William Pratt House
9 Dewey Court
Northampton
Massachusetts 01060
USA

A catalogue record for this book
is available from the British Library

Library of Congress Control Number: 2021947983

This book is available electronically in the **Elgar**online
Political Science and Public Policy subject collection
http://dx.doi.org/10.4337/9781802203110

ISBN 978 1 80220 310 3 (cased)
ISBN 978 1 80220 311 0 (eBook)
Printed and bound by CPI Group (UK) Ltd, Croydon, CR0 4YY

Contents

Preface

Many years ago, in the early 2000s, I worked with children and young people's use of new online technologies. Back then, the average adult population mainly knew the internet as a communications platform for emails, basic search and news. Younger generations, on the other hand, were quickly adopting the internet's opportunities for self-expression and social networking as a digital extension of their everyday life. It was an online world for youth perceived by an adult generation as inaccessible, incomprehensible and secret. These were also the early years of the popularisation of an online privacy movement. Although it had existed in technical activist communities since the introduction of the World Wide Web in the 1990s, the privacy movement was increasingly also sieving into public awareness. Online privacy was considered a form of power that we could campaign for and tell citizens to demand by using privacy-enhancing technologies to protect ourselves against state and commercial tracking and surveillance. At one point I realised though that educating and raising awareness among the users of the internet was not enough. I became particularly concerned that youth's initial experience of freedom away from adults' prying eyes was actually just another form of control by other more invisible powerful actors, such as social media tech giants. What concerned me the most was these actors' presence everywhere – at our events and meetings, in public consultations, in policy initiatives. It was as if their business design and model for the evolution of the Internet was the only formula possible. Therefore, I started focusing on alternatives to the very design and business models of these services inspired by the early critical voices in the field.

All along, civil society actors had worked to have human rights issues included on the official internet governance agenda in processes such as the World Summit on the Information Society (WSIS) and the UN Internet Governance Fora (IGF). However, it was not until 2013 that the United Nations General Assembly affirmed that the same rights that people have offline must also be protected online (UN, 2013). Even then, in the more general business and public discourse, human rights online did not take a proper foothold. Human rights considerations, such as concerns regarding our rights to privacy and data protection, were often in public discourse described as obstacles to digital innovation, as old social norms that were preventing an unavoidable digital evolution of society. It was obvious that the very business and technology culture of the internet's development was preventing a more ethically reflective and constructive debate. In 2014, I therefore established the Global Privacy as Innovation Network at the UN Internet Governance Forum, bringing together industry, human rights advocates and technology entrepreneurs to explore privacy as an opportunity rather than an obstacle.

At that time, data ethics was still not a layman's term and the ethical implications of data technology and business were addressed in public debate – if addressed at all – in terms of privacy implications only. As such, it was still a great struggle for the lone privacy activist to walk into a public debate on social media and the digitalisation of society. Human rights issues of online business were considered an activist topic separate from the debates on big data innovation and disruption that were shaping online business development.

In 2015, I left the Danish Media Council where I had worked for 10 years in a national EU Awareness Centre on youth's use of the internet and new technologies. Together with the former journalist Pernille Tranberg, I started exploring a growing movement among technology designers and emerging companies developing and promoting alternative data design and business models based on the preservation of privacy. We established the thinktank DataEthics together with two other women. At first it was like fighting with the most popular kid in school; we were the outsiders, the activists, who did not understand the awesomeness of this reckless kid and his shiny new tools. However, public discourse and awareness was also changing; in particular, the negotiations of the European General Data Protection Reform were increasingly addressed in public debate. Alongside a growing awareness and attention to the ethics of digital technology, the thinkdotank DataEthics became increasingly involved in the public debate as well as in business and the policy debate.

These years of immersion in the field have been integral to my understanding of the history and power dynamics that I explore in this book. I have seen a rising awareness of the role of the internet in society that increasingly included a view on risks and challenges beyond mere technical and functional issues. And I have seen how a focus on online privacy risks transformed into a more generally accepted awareness of data ethical implications and issues in the general public – violations of human dignity, the exposed vulnerable groups in society, discrimination and challenges to democracy and democratic institutions. This is also where I discovered a pattern of cultural powers; the distribution of power in culturally embedded socio-technical data systems, and not least the interests invested in the very term data ethics as an expression of power dynamics.

Acknowledgements

Many people and communities have been vital to this book, and I want to thank them all – the privacy and human rights advocates, the whistle blowers and investigative journalists, the open-minded policymakers, the brilliant scholars. Specifically, I wish to sincerely thank the people with whom I have had the most empowering collaborations and exchange of ideas over the years: Francesco Lapenta, Pernille Tranberg, Birgitte Kofod Olsen, Rikke Frank Jørgensen and Carolina Aguerre. A special thanks to Jens-Erik Mai for the invaluable reading of and comments on several drafts of what turned into this book.

Thank you to Klaus Bruhn Jensen, Simone van der Hof and Safiya Umoja Noble for their insightful reading and feedback and to the research group 'Surveillance, Information Ethics and Privacy' at the University of Copenhagen – in specific Laura Skouvig, Karen Søilen and Sille Obelitz Søe. I also would like to extend my gratitude and appreciation to the people behind the IEEE Ethically Aligned Design Initiative who put ethics into action; the human rights scholars and advocates of the global internet governance community who paved the way – in particular, Meryem Marzouki and Marianne Franklin; the members of the Global Privacy and Innovation Network who all very early could see privacy as something other than an obstacle; the inspiring Data Pollution & Power group, and in specific Aimee van Wynsberghe, at University of Bonn's Sustainable AI Lab; Gianluca Misuraca of the InTouchAI.eu - International Outreach for a Human-Centric Approach to AI initiative; the trustworthy AI civil society movement – in particular Mark Surman and Martin Tisne; Nathalie Smuha and the EU High-Level Expert Group on AI who coined the term 'Trustworthy AI' and revitalised the human-centric approach.

I am forever thankful for the support of close friends: My guardian spirit Britt, Carolina, Signe, Stine, Rikke, Marie, Maria, Camse. Grateful for family love now, then, here and beyond: Francesco & Clara, Bo, Ask, Sanne (and the three), Virginia & Francesco, Maurizio, Tina & Mario, nonina, Mino, Luca & Lucia (and the Rome family), Pino & Franca (and the Bergamo family), Nicola & Franca, Virginio, Rocco, Vincenzo, Gry & Jacub (and the Sletten family), farmor & farfar, mormor & morfar, Susie, mor.

Introduction to *Data Ethics of Power*

'Imagine a piece of music which expresses love. It is not love for any particular person. Another piece of music will express another love. Here we have two distinct emotional atmospheres, two different fragrances, and in both cases the quality of love will depend upon its essence and not upon its object. Nevertheless, it is hard to conceive a love which is, so to speak, at work, and yet applies to nothing.'
Henri Bergson, 1932

Why do we need to talk about data ethics? The relation between humans and their data and information has always involved social and ethical dilemmas, and it is no novelty that data systems and registers have persistently throughout history reinforced power dynamics in society and also created new ones. Nevertheless, at the beginning of the 21st century the very shape of our contemporary data systems, their ethical implications and power complexes have rapidly transformed with the advancement of new digital technologies and data, which is why I, in this book, propose that we also need a differently shaped data ethics. A recent and new development is the transformation of all things into data as an effortless, costless and seamless extra layer of life and society. Data at the time of writing no longer just captures politics, the economy, culture and lives – data is their extension. It is ingrained in society in multiple forms in increasingly complex digital systems which have been developed to contain and make sense of large amounts of data and to act on that knowledge. These digital data systems form a key component of decision-making in politics, culture and industries, and also on life trajectories, and consequently, they are also the centre of power negotiations between different interests. As such, this new shape of power should be at the core of any ethical concern with data systems – and at the heart of data ethics.

With this book, I want to find a common ground for debates on the development and status of big data and AI sociotechnical environments by spelling out a 'human approach', which I refer to as a 'data ethics of power'. A data ethics of power is concerned with making visible the power relations embedded in big data and AI sociotechnical infrastructures in order to point to design, business, policy, social and cultural processes that support a human-centric distribution of power. But what does this actually mean? Imagine an AI robot that sieves through pictures containing predominantly white faces deciding what human 'beauty' means. Ponder on an online search system that learns from news articles to recognise words such as 'nanny' and 'receptionist' as female and words such as 'architect' and 'financier' as male. Consider an AI assessment model that scores an otherwise brilliant teacher badly because it cannot interpret the social and human dimension of the teacher's work. Reflect on a mass surveillance global intelligence network or a mass political profiling campaign enabled invisibly by social networking sites with the personal data of millions of people. These sociotechnical data systems and practices are ethically questionable. Most often they are 'unfair', certainly they are not morally 'good' but 'bad', and in some contexts they might even be deemed illegal. However, in this book I propose that data ethics is not a legal assessment, neither is it a moral evaluation of the good or the bad. As said, a data ethics of power wants to expose the power relations embedded in big data and AI sociotechnical infrastructures. It aspires to find ways to build things, act and govern in a holistic manner that benefits human societies, or what, with reference to the French philosopher Henri Bergson, I seek to advocate as open love. Love in the most literal and figurative sense of the word. Love for humanity. Love for the planet and its inhabitants. Love for each other. Love for the universe and beyond. Love without a specific interest. Just love. Nothing more and nothing less.

An encouraging turn in the early 21st century is that the conversation on data ethics has matured into a main topic of public debate and a crucial policy agenda item in Europe and beyond. Remarkably, as I will illustrate throughout the book, an ethical stance on data and data innovation, particularly in the late 2010s, transformed into what was considered Europe's competitive position on a global geopolitical stage. A range of societal stakeholders from industry, academia and civil society to governments and intergovernmental organisations were here presenting and trying to solve the various critical problems in our current data infrastructures: data bias and a lack of multicultural representation, the disempowerment of consumers, democracy challenged by black box algorithms, flawed IT security and data protection, data monopolisation, voter manipulation and many more. Furthermore, the lost opportunities of locked-in data for science, public service, business and, in particular, the technical development of AI systems were presented as major problems in the state of affairs.

The problems were then, and are always in certain critical moments, evident. But how *do* we then change the direction of a sociotechnical development that we have identified to move contrary to our human values, challenging human dignity, cohesion, agency, responsibility and democracy? Different stakeholders identify different problems and equally diverse solutions are suggested. And in this moment interests are of course bound to clash, because not all problems with our data reality converge and neither do their solutions. This is the moment when real controversy arises, when power takes material form. I propose that to achieve change in the age of big data and AI, we need to take an extra look at this moment, because it is in this controversial moment – when critical problems become visible, when value and interest negotiation take centre stage and compromises are made – that actual sociotechnical transformations emerge.

At times it seems too overwhelming – the complexity of issues, challenges and the compromises we have to make. Though, as I will argue in the book, not all is lost to the state of affairs. The book is built on the idea that humans are in control of these compromises – but only if we see the entire field of powers and interests invested in a sociotechnical transformation, including our own. The greatest challenge we have today is ourselves. What I see in the 'governance' efforts of policy and decision-makers in state, intergovernmental, technical, industry and civil society communities is that when attempting to orchestrate the symphony of sociotechnical acts and agents that comprise sociotechnical change, we repeatedly fail to address its complexity. Our vision is clouded by our particular problems and brilliant solutions, and therefore, we do not see the core problem of sociotechnical change, which is the very shape of the power dynamics that effect change and outline the direction of sociotechnical development. We do it over and over again. We undervalue the complexity of the powers embedded in our sociotechnical environments and the multifarious ethical and social implications. We neglect to look beyond our own interest, fields of application and expertise, and because of this we create implications for ourselves, and even produce new ethical problems. Time after time, we overvalue our own – although predominantly well-meaning, but also often very limited – perspectives on problems and solutions. We fail to coordinate, to translate amongst each other and, most crucially, to assess broader social and ethical implications. I will therefore in this book present a point of reference, where we can meet, by spelling out the human approach – a data ethics of power.

The book contributes to debates on data ethics, big data and Artificial Intelligence in three ways:

1.   A Data Ethics of Power

'Data ethics' at first gained traction in public discourse as a term to address the general socially and ethically problematic sides of big data technologies and systems in addition to their challenges to privacy. Then data ethics came to represent the 'good intention' of primarily companies and states and it became part of an 'ethics by design' applied ethics, a moral philosophy, with methodologies and practices designed to instil good human values into big data and AI systems. In the book, I part from a concern with the morally 'good' and 'bad' of big data and AI and I position data ethics as a concern with the cultural and social power dynamics of big data and AI. Technologies are cultural products and technological practice is embedded in socially ordering cultural systems of meaning-making that are lived and experienced by individuals. This is why cultural systems of sociotechnical change and practice per se are relevant as ethical problems that we should seek to solve with an applied data ethics of power. I argue that we need to be ethically concerned with the constitution of big data and AI infrastructures as cultural systems of a type of social ordering, in which interests of dominant actors in society have the primary advantage while other minority interests are further disadvantaged. Sociotechnical digital data systems are spatial architectures that reinforce and distribute power. They have data cultures that sustain power for some while repressing the freedom and agency of others, and they are the locus of different powerful interests: corporate, governmental and even scientific.

2.   Data Ethical Governance

Being actively involved in the evolving public debates and policy debates on the ethical implications of emerging data technologies, I also seek to provide an explanation as to why data ethics became the centre of these discussions. By asking and answering this question, I at the same time outline a governance role and function for data ethics in the context of sociotechnical change. As such, I also propose a common ground for the debates and negotiations that I am involved in based on a human morality and approach to the power structures of big data and AI sociotechnical systems.

I want to illustrate what data ethics can do for human governance in the context of sociotechnical change. I examine the public debates and policy agendas of the late 2010s on the infrastructures of the Big Data Society as one 'in-between' phase of the general phases of the sociotechnical development of a large technological system (Hughes, 1983, 1987) in which different technological cultures and approaches compete to gain technological momentum. I argue that human governance of big data and AI adoption and developments entail a critical awareness of the ethical compromises we make in these moment of controversy, as they constitute the cultural compromises invested in the big data and AI sociotechnical infrastructures' technological momentum for consolidation in society. Accordingly, the vigilance of the cultural powers invested in this is vital. I here urge that we enrich a culture of human power in the governance, design and adoption of big data and AI sociotechnical infrastructures.

3.   The Human Approach

The 'human-centric approach' is a term and a theme that has emerged, particularly in global policy discourses on AI, with no common conceptualisation other than an emphasis on the special role and status of humans. We could, and it has been done, argue against such a human-centric approach and its assumed anthropocentricism if it was primarily concerned with the individual human being and the human species as such. However, in the book I propose that we look at the concept in an entirely different way and I provide a supplementary conceptualisation of the human-centric approach, what I call a 'human approach'. The human approach, I suggest, is one concerned with the role of the human as an ethical being with a corresponding ethical responsibility; or in other words, the human approach is not about prioritising the individual human being (it is not about 'individualisation', as Zygmunt Bauman (2000) would have called it) – it is about the human as an ethical being, our human *ethical responsibility* for not only ourselves but for life and being in general, and it is about prioritising the *human dynamic qualities, a human infrastructure of empowerment*, in very concrete ways in big data and AI sociotechnical infrastructures. That is, the human approach also encourages, in practical terms, the empowerment of dynamic human moments in their very data design, use and implementation, which does indeed also include, but is not limited to, the empowerment of the individual human being.

A human approach is best expressed in what I refer to as data ethics 'spaces of negotiation' and 'critical cultural moments'. The spaces of negotiation enable critique and negotiation, but they are only possible when 'systems' (material, immaterial, technological, cultural etc.) clash and controversy arises. The critical cultural moments have special human characteristics and are possible when human memory and intuition are privileged and provided time and space to tinker. In practical terms this is expressed in a prioritisation of the human interest in the data of big data and AI infrastructures via the meaningful involvement of human actors in their very data design, use, governance and implementation.

Thus, the approach of a data ethics of power is not just about humans –it *is* human, which essentially means that it cannot be put aside; neither can it just be applied when considered useful. We need to think of data ethics as a human morality rather than just a social morality (Bergson, 1932/1977). We can formulate data ethics guidelines, principles and strategies, and we can even program artificial agents to act according to moral rules. However, to truly ensure a human-centric distribution of power, data ethics must become more than just a moral obligation, a set of programmed rules. It must be human.

The philosopher Henri Bergson, who I will refer to extensively in the last part of the book, provided in his philosophical writings an excellent illustration of what this means. It does not mean that humans are bestowed with divine gifts. It does not mean that we are non-natural extra earthly beings. All it means is that humans do not only have the same intellect as machines. We have an additional philosophical capacity which is an intuitive one (Bergson, 1896/1991; Deleuze, 1966/1991), the capacity to 'think movement' (Bergson,

1907/2001, p. 318) by setting into motion our memory (Bergson, 1896/1991). Most importantly, he raised an essential critique of utilitarian approaches to the living, and by doing so, I would argue that he simultaneously provided us with a conceptual map to also understand the limits of the intellectual capacities of both humans and AI, but that only AI cannot exceed.

Finally, what can we achieve with this human approach? With a human approach we counter the closed and exclusive properties of big data and AI systems that sustain a lived reality of control and order of exclusive societies. These sociotechnical systems materialise interests and enact power asymmetries in society. They only represent a slice of a dynamic moving human reality and multiple cultures, yet they act, and are increasingly also adopted as if they were complete. A human approach counters these exclusive tendencies with love. Love is a concept found in all cultures, from the Greek term, *agape*, signifying unconditional love for all of humanity to the Buddhist term, *maitri*, expressive of universal loving-kindness. In this book, I use Henri Bergson's concept of open universal 'love' that has no interest, but is universally directed at the whole of humanity and that therefore enables an open and just society (Bergson, 1932/1977).

OUTLINE OF THE BOOK

I present three core competing structures of power, each with their shape and style – and each with their 'data cultures'. The book consists of three parts, emphasising the following three different characters of power:

1. On power and big data (Chapters 1 and 2)
2. On power and AI (Chapters 3 and 4)
3. On human power and data ethics (Chapter 5)

In the first two chapters of the book, on *power and big data*, I address the first kind of power, big data sociotechnical infrastructures (BDSTIs) and their cultures and environments. BDSTIs are constituted physically with fibre cables that run across the globe, enabling data collection and access across geographic territories and jurisdictions, and virtually in spaces of flows around which dominant societal functions are increasingly organised (Castells, 2010). BDSTIs also constitute a redistribution of power facilitated by these new technologically mediated configurations of space and time. To design and shape the infrastructural components of BDSTIs is here an essential form of power, and, for example, surveillance powers of state and industry actors are embedded in BDSTIs as a key property of their architecture and design (Haggerty & Ericson, 2000; Lyon, 2001, 2010, 2014, 2018; Hayes, 2012; Bauman & Lyon, 2013; Galic et al., 2017, Clarke, 2018, etc.).

In Chapters 3 and 4, which constitute the second part of the book, on *power and AI*, I focus on the type of power that is concentrated in the emergence of big data artificial intelligence sociotechnical infrastructures (AISTIs). These are, first and foremost, evolutions of the analytical capabilities of BDSTIs, and constituted as BDSTIs, but with components designed to sense the environment in real time, learning and evolving with autonomous or semiautonomous agency.

The two types of sociotechnical infrastructure described in the first four chapters of the book constitute two forms of power that work in different dimensions of human reality and society. While BDSTIs primarily act in our space by transforming all into immobilised digital data, AISTIs also occupy our time by acting on that data to actively shape the past and present in the image of the future. Accordingly, I propose that a core concern of a data ethics of power should be with AISTIs' and BDSTIs' constitution as cultural systems of a type of social ordering, in which interests of dominant actors in society are spatialised and immobilised and thus more difficult to criticise and renegotiate.

Finally, in the last chapter and part of the book I discuss *human power and empowerment*, which I claim is at the core of a data ethics of power. The human approach of a data ethics of power concerns the role of the human as an ethical being with a corresponding ethical responsibility. However, ethical agency requires special spatial and temporal conditions to flourish. Human power is, in the face of sociotechnical changes, in constant negotiation with the power of BDSTIs and AISTIs. This therefore requires an applied data ethics that ensures the involvement of human actors in the very data design, governance, use and implementation of sociotechnical data systems.

HOW TO ATTACK A COMPLEX DATA ETHICS PROBLEM

When I started writing this book, I had already been working in the policy and practice field for many years. Data ethics had become a layman's term and critical showdowns with powerful tech giants were finally acceptable in the public debate. Still, even when ethical reflection and social awareness were present, I saw how we often failed to assess across our different cultures of interests the implications of what we did. As described before, we are immersed socially and technologically in sociotechnical and cultural structures of power that limit us in what we do and what we think we can do with technology.

Normally, when scientists approach a problem, the most traditional way to do this is to look at the problem from one specific area of expertise (such as ethnography, law, philosophy, sociology or engineering). What I have realised is that these separate fields are actually also characterising the way people are talking about AI, big data and data ethics in public discourse and in policy in general. I have, over a very long period, been exposed to the different languages and traditions of the stakeholders and scientific communities that shape this field, which has educated me to see the idiosyncrasies and language of each group and tradition: how they deferred to one another, but also how they touched upon each other – that is, what they had in common.

What I have tried to do in the book is to integrate different fields into a more organic way to approach a problem. I wanted to create a synthesis of these languages, integrating different traditions, to represent the common voices I have been interacting with. And it has of course been an incredibly complex labour of relating and reporting on the different traditions and synthesising them in a meaningful way. What I have basically done is to try to embrace the complexity of sociotechnical change, which is in no way an easy task.

How do we create a meta-language where, specifically in technology development, you have fundamentally different, sometimes even contradicting, traditions that have to talk about the same problem?

Here, I was particularly inspired by Thomas J. Misa (1988, 1992, 2009), who addresses this complexity with an approach that moves between different levels of analysis. He suggests a 'multi-level' analysis that encompasses both micro and macro perspectives, which he does to overcome what he considers a false dichotomy between two different framings: the relation between humans and machines produced by analyses that adhere strictly to either a micro or macro perspective on technology (Misa, 1988, 2009). Focusing on either the micro-dynamics of, for example, designers and engineers of a technology or, on the other hand, only on larger macroeconomic or ideological patterns, will present very different and often conflicting views on sociotechnical change, he argues. That is, while the former will often not see the wider social conditions and power dynamics for change, the latter will just as often reduce individual nuances and factors by making sense of them in terms of larger societal dynamics only. A multi-scalar analysis, conversely, acknowledges both perspectives.

Three scales of times (micro, meso and macro) are also central to my delineation of a data ethics of power:

On the micro scale, the very design of a technology can be discerned as closed or open to human controversy and negotiation. For example, an AI agent's algorithms and data processing can be a black box and evolve autonomously without human intervention. Alternatively, it could have a human in the loop design, transparency of design, auditability, and personal data control. Focusing on the design of the technology, I focus on the micro time and space in which it is designed and programmed by human designers or when it is implemented by human users. The micro-scale temporal analysis of a data technology here considers whether the very data design and the design process for this are open for cultural value negotiation. For example: Is the design process locked in an unquestioned technological data culture? Or is it open for critical assessment in terms of the invested interests in the process?

On the meso scale, institutions, companies, governments and intergovernmental organisations will be open or closed to negotiation of the values and cultural frameworks for their practices. Closed entities will move along the natural state of affairs and seek only basic compliance with law. Open entities will establish initiatives and practices dedicated to value negotiation and ethical reflection in addition to legal compliance. When patterns in ethical reflection on a specific topic are identified across initiatives and entities, 'data ethical governance' may be introduced.

Lastly, sociotechnical developments and change can be analysed on a macro scale of time. Moments of ethical reflection and negotiation here emerge in between crisis and consolidation of a sociotechnical system in society (Hughes, 1983, 1987; Moor, 1985). These moments are critical as they constitute social negotiation and result in cultural compromises, namely, 'the technological momentum' (Hughes, 1983, 1987) that a sociotechnical system needs to evolve. They are also crucial to phases of innovation and development as they constitute the transformation of the sociotechnical system that emerges from a quest to solve the critical problems of the system. We can here think of 'a battle of the systems' in which an old and a new system exist simultaneously in a relationship of 'dialectical tension' (Hughes, 1983, p. 106–39). Or a moment of conflict and resolution not only among engineers but also in politics and polymaking. In these moments of conflict, critical problems are exposed, different interests are negotiated, and they are finally gathered around solutions to direct the evolution of the system. The new system, or the transformation of the old system, evolves out of the problems identified and solved in this phase. These critical problems of the system are not just resolved as technical problems, for example, with the agreement on technical standards with systems requirements, but are in dialogue with political and historical factors.

THE TERMINOLOGY

If we want to talk about the same thing, we also need a shared terminology. A key objective of this book is to create a common ground for data ethics debates on AI and data with a focus on human power. Part of this has been to develop a common discourse for a data ethics of power with some key terms that I introduce here.

Data Ethics

Data ethics addresses the distribution of power and power relations in the Big Data Society and the conditions of their negotiation and distribution. Applied data ethics is concerned with making these power relations visible in order to point to design, business, policy, social and cultural processes that support a human(-centric) distribution of power.

Power

Power is a well-known, contested and theorised concept with roots in various disciplines and sciences, from Hobbes to Marx, Arendt and Foucault and many more. Here, I only explore the latest peak of contemporary theories on power that specifically address contemporary cultures and digital technology. A data ethics of power is concerned with the distribution of power facilitated within new technologically mediated configurations of space and time (or what I call 'BDSTIs' and 'AISTIs'). It recognises value and interest cultural power struggles and negotiations as a core component of sociotechnical change and governance. This conceptualisation of power and technology stems most profoundly from surveillance and critical data studies that address the state of power in the Big Data Society on the level of the micro-design of systems, business, state and engineering data practices and on the level of macro-societal sociotechnical change. In particular, we need to address power in terms of its 'liquidity' (Bauman, 2000; Bauman & Haugard, 2008; Lyon, 2010; Bauman & Lyon, 2013). That is, we should concern ourselves with a type of power that is concentrated and engineered by a few power actors, yet also increasingly self-sustained, re-engineered and evolving in (surveillance) cultures (Lyon, 2018) of use, design, governance and imagination, and therefore difficult – but not impossible – to change. Conceptualisations on culture and power I derive from a tradition within cultural studies that addresses the uneven distribution of power in cultural representation, cultural practices and products. Crucially, in this perspective, cultural power is never stable and can always be challenged and redistributed.

Sociotechnical

Technology is always part of society, just like society is always part of technology. This also means that one cannot understand one without the other. Technology is not only design and material appearance but also *sociotechnical*; that is, a complex process constituted by diverse social, political, economic, cultural and technological factors (Hughes, 1987, 1983; Bijker et al., 1987; Misa, 1988, 1992, 2009; Bijker & Law, 1992; Edwards, 2002; Harvey et al., 2017 etc.)

(Sociotechnical) Infrastructures

Infrastructures are the virtual and material sociotechnical organisations of the space of societies. They are engineered and directed, but they also evolve in social, economic, political and historical nonengineered dynamic contexts. Specifically, a sociotechnical infrastructure is a particular type of human-made space which is the material and immaterial, engineered and nonengineered processes that evolve in a space of negotiation and struggle between different societal interests, imaginations and aspirations (Star & Bowker, 2006; Bowker et al., 2010; Harvey et al., 2017 etc.).

BDSTIs (Big Data Socio-Technical Infrastructures)

In the first part of the book, I introduce the term BDSTIs to refer to sociotechnical infrastructures constituted by big data technologies. They are the primary infrastructures of the flows (Castells, 2010) of global economies and societies cutting across geographic territories, legal jurisdictions and cultures. In the early 21st century, BDSTIs were increasingly representing and constituting global societies and environments as the mundane background against which social practice, social networking, identity construction, economy, culture and politics were conducted. They were in part institutionalised, in systems requirements standards for information technology (IT) practices, and in regulatory frameworks for data protection, and they were invested with human imagination about the challenges and opportunities of big data.

AISTIs (Big Data Artificial Intelligence Socio-Technical Infrastructures)

AISTIs is a term I use to describe an evolution of the analytical capabilities of BDSTIs. AISTIs are constituted as BDSTIs but with components designed to sense the environment in real time, learning and evolving with autonomous or semiautonomous agency. While BDSTIs act in space by transforming all into immobilised digital data, AISTIs also occupy time by acting on that data to actively shape the past and present in the image of the future. In the second part of the book, I focus on the history, ethics and development of AISTIs.

Culture

There are two sides to culture: (1) It is a system that brings together communities with shared conceptual frameworks and resources, and it is an active system with specific priorities, goals and ways of organising the world that are actively imposed in society. (2) Culture is 'a whole way of life' (Williams, 1958/1993). It does indeed consist of prescribed dominant meanings, but importantly culture is also the negotiations of these meanings. That is, culture is not just one. It is multifaceted – institutionalised and formalised, and practiced by dominant groups in society – but it is also subcultural and practiced by, for example, minority groups in society. Thus, culture is never stable; it is from the outset a constructed system of meaning-making and is therefore also always up for contestation and social negotiation.

Data Cultures

The cultures that frame data science, practice and governance we may call 'data cultures'. They are culturally coded conceptual maps of the engineers, data scientists and designers of data systems; deployers of data systems; legislators of data systems; and users of data systems. They are not always shared, and they may even be in conflict. Data cultures are interrelated with societal power negotiation and struggle. The very practices of data scientists and designers are, for example, framed within specific informal or institutionalised cultural systems of meaning-making. Accordingly, the very practice of developing a data system and design is a cultural practice.

The Human Approach

The 'human-centric' or 'human-centred' approach was a popular term in late 2010s' policy and advocacy discourses on the ethics of AI and big data, used as a way to recentre the sociotechnical developments in these fields on the human interest. In this book, I further explore and conceptualise this term, but I refer to it as the 'human approach'. I do this to emphasise the role of the human as an ethical being with a corresponding ethical responsibility for not only the human living being but also for life and being in general. In practical terms the human approach is associated with the human interest in the data of AI through the involvement of human actors in the very data design, use and implementation of AI. The human approach of a data ethics of power specifically constitutes a critical reflection on the power of technological progress as well as the big data and AI sociotechnical systems we build and imagine.

Values and (Data) Interests

Values are 'idealised qualities or conditions in the world that people find good' (Brey, 2010, p. 46). They are represented in power struggles over different cultures and worldviews. Interests are held by different actors and represent social power struggles over, for example, material things such as resources. A data ethics of power is particularly concerned with the interests in data invested in data design and governance ('data interests', Hasselbalch, 2021). Values and interests are core components of sociotechnical change.

Moral Agent/Agency, Ethical Agent/Agency

The concept moral agency is often used interchangeably with ethical agency. However, I want to make a distinction between the two to emphasise the difference between two different capacities. I understand a moral agent as one that can only enforce and act according to moral prescription and determination. For example, 'intelligent' nonhuman agents (AI agents) are moral agents, but they are not ethical beings. This is also why I consider data ethics a human responsibility only.

Human Actor/Agent and Nonhuman Actor/Agent

I deliberately make a distinction between human and nonhuman agents and actors. However, my intention with this very rough distinction between the human and nonhuman is not grounded in a technological nor cultural determinism; rather, it is a semantic trick aimed at disclosing the limits of the moral agency of AISTIs in particular and, in this connection, the importance of human ethical agency and power to change and govern sociotechnical development. Accordingly, despite my distinction between the two, I do recognise technological artefacts as extensions of human agency and intent as well as their increasing indistinguishability within human environments.

Ethical/Data Ethical Governance

'Ethical governance' (Rainey & Goujon, 2011; Winfield and Jirotka, 2018) is a multi-actor, reflexive, open-ended (Harvey et al., 2017; Hoffman et al., 2017) and agile process designed to ensure the 'highest standards of behaviour' (Winfield and Jirotka, 2018). It goes beyond just good and effective governance. I define 'data ethical governance' as a form of ethical governance that specifically addresses the complexity of the Big Data Society with infrastructural practices that create human-centric data cultures.

1. Big Data Socio-Technical Infrastructures (BDSTIs)

'Change: survive!'
John Mashey, 1999

From 1967 to 2018, Morandi Bridge ran as one of the main arteries through Genoa, Italy, connecting the east to the west part of the city. The bridge's concrete construction was a global symbol of Italian engineering and technical capacity. In fact, Italy was in 2018 one of the top cement producers in the world and thousands of concrete viaducts, tunnels and bridges worldwide were based on Italian design. As a key infrastructural component of the city, Morandi Bridge formed the silent background against which life and business were facilitated with thousands of people driving across it every day without giving the bridge and its construction an extra thought. However, on 14 August 2018, the heavily trafficked bridge collapsed, causing the death of 43 people and leaving 600 homeless. The bridge was at once no longer silent. With countless media reports and investigations, not only was the engineering history of the bridge told but the very infrastructural breakdown was equated with the collapse of a 'national myth' (Mattioli, 2019).

Think of Big Data Socio-Technical Infrastructures (BDSTIs) as being like Morandi Bridge, or just any ordinary road or building that resides in a 'naturalised background' (Edwards, 2002, p. 185). We cross them, like we cross bridges and follow roads, every day. Unnoticed, they facilitate and organise our everyday lives. They constitute the micro-spatial architecture of our lives and they are embedded in macro-societal structures. And just like Morandi Bridge, these BDSTIs are not just appearance, digits and cables; they have a politics and a culture that become particularly apparent in moments of crisis and infrastructural breakdown (Star & Bowker, 2006; Bowker et al., 2010; Harvey et al., 2017).

In this chapter, we will examine the infrastructures of the Big Data Society with the double purpose picked up from Infrastructure Studies: to know and make transparent a human environment, but also, crucially, to control it (Harvey et al., 2017, p. 2). The main objective is to understand the special power dynamics of the sociotechnical infrastructures that a data ethics of power addresses. I refer to these as BDSTIs, or AISTIs, and investigate how they have evolved as components of the power dynamics of the Information Society, addressing specifically the conception of a European big data infrastructure. In the last part of the chapter, we will look at the ethical problems that concern a data ethics of power specific to the Big Data Society, which is the term I have chosen to use to describe the specific characteristics of societies in which BDSTIs and AISTIs are dominant. We will also explore the asymmetric experience of data power and examine the voice of a data ethics of power. Who has the power to raise issues, define problems, or propose and create the solutions to the problems we are facing?

1.   BIG DATA SOCIETY

What is a Big Data Society? Society's technologically advanced big data technologies and systems aside, how do we imagine the role of big data in a society like this? What social, economic and cultural functions should it have? If we consider the Big Data Society a coherent social structure, as a specific complex of powers with equally particular critical ethical problems, we also need to understand its social, economic and, not least, ideological underpinnings. Professor of internet governance Viktor Mayer-Schönberger and journalist Kenneth Cukier depict the Big Data Society as a societal revolution that transforms human work, social relations and the economy. This is a transformation enabled by computer technologies and dictated by a transformation of all things (and people) into data formats ('datafication', Mayer-Schönberger & Cukier, 2013, p. 15) in order to 'quantify the world', thus helping businesses, governments and scientists organise and make sense of data (ibid., p. 79).

This evolution of BDSTIs can be coupled with the imagination of big data as an unlimited resource that, unlike other resources in society (e.g., oil and water), will not diminish (ibid., p. 101). In essence, we could therefore also argue that big data more than anything is also a movement behind which lies a system of imagining and making sense of the role of digitalised data in society (Mai, 2019, p. 111). The collection of big data is here perceived as an end in itself, holding the promise of future endless ways of use and reuse (Mayer-Schönberger & Cukier, 2013, p. 100).

Ideas about the risks and potentials of big data can be traced back to the late 1990s, when big data surfaced as a term in the computer science and business fields to describe a range of emerging technological innovations in digital data storage, exchange and analytics enabled by computer technologies and the evolution of the internet. It was first used by a chief scientist named John Mashey at Silicon Graphics, a large US-based computer graphics company, in a number of product pitch talks depicting the great promise of big data, but also describing the commercial and technical challenge to meet this future potential (Lohr, 2013). For example, in 1999 he predicted how big data would unsettle both human and material IT infrastructures. The response was an urgent call for companies to 'Change: survive!', as one of his PowerPoint presentations, with the title 'Big Data and the Next Wave of InfraStress', exclaimed in 1999, with reference to, among others, enhanced computer power to store and process data and the unleashing of data with scalable interconnect and high-performance networking (Mashey, 1999, p. 45).

The limits and risks are within a big data movement mainly perceived as 'technical' in nature: limited storage, processing and analytical capacities. The most powerful companies and institutions are those with a 'big data mindset', engaging in big data infrastructural practices, collecting big data, and processing and creating interoperable big data sets (Mayer-Schönberger & Cukier, 2013, p. 129). Success equals the ability to transgress the limits and borders that lock in valuable data. Data retained and data potentials locked in space are identified as the problems to solve. First the problems and their solutions are technical, as in, how do we store and collect data? How do we make sure that we have tools to make sense of it all? Then they become legal, as in, is there a way to comply with a data protection law and still collect and process as much data as possible? Finally, they become human. How do we retrieve data beyond the limits of individual rights like privacy, beyond the human body, the mind, the human inexplicable nature? The limits and borders crossed to retrieve data are increasingly human and, in this way, a big data movement becomes a human challenge.

BDSTIs are human-made spaces shaped by human imagination, compromise and domination of some interests over others. A driving force here has been the commercial and institutional fantasies about the potential of big data as an unlimited resource and the commercial and technical risks to companies and infrastructures that fail to store, collect and process it. BDSTIs are spaces that are therefore also shaped by practices aimed at making the most of the promise of big data while simultaneously mitigating identified challenges to its fulfilment.

The Imagination and Politics of Space

Think about 'unprocessed' spaces of the past. Hundreds of years ago, an untouched breadth of land cut from the Rocky Mountains, in the west of North America, across to Illinois and Indiana in the east and extending from Canada in the north to Texas in the south. About this land, which was named the 'prairies' ('the grasslands') by the French first settlers, the American poet, Emily Dickinson, in the mid 1800s dictated a small poem to a friend: 'To make a prairie it takes a clover and one bee, One clover, and a bee, And revery. The revery alone will do, If bees are few' (Franklin, R.W., 1998). Even space untouched by human hands, she seemed to say, could be captured and made meaningful by human imagination. In the 21st century, nothing is left untouched, very little space is open and plain, but rather predominantly consists of human-made architecture – material and digital. Very rarely do we sit and ponder over open plains. We move through buildings, parks and playgrounds, among planted trees and bushes – and through digital networks. All of which are invested with human interests, intentions and imagination. Yet, there are still plains left to explore. Imagination does not stop at the break of the horizon. As Rebecca Moore, Director of Google Earth, Google Earth Engine and Google Earth Outreach said with great enthusiasm in 2015, 'Just imagine the next generation, it's like a living breathing dashboard for the planet' (Eadicicco, April 15 2016).

In 1974, the Marxist philosopher and sociologist Henri Lefebvre defined space as an architecture that we can touch, feel and be bounded or liberated by with our bodies, but also with our social realities and minds. Space is, he argued, a composite of a material physical reality and social practice, or a type of space that does not exist without 'the energy that is deployed within it' (Lefebvre, 1974/1992, p. 13). He divided this social energy invested in space into three types: 'the perceived, the conceived, and the lived' (Lefebvre, 1974/1992, p. 39). In other words, we perceive space physically with our perception, and we feel space qua our positioned bodies; however, space is also *conceived* by, for example, urban planners, engineers and scientists and it is lived with imagination that seeks to 'change and appropriate' it (Lefebvre, 1974/1992, p. 31). In this way, he pointed to struggles over the meaning of space, delineating the power dynamics and politics that shape space as a real and imagined resource invested with specific interests. Generally, space is open for active 'occupation' by interests, and evidently our 'global space' was originally also nothing more than 'a void waiting to be filled, as a medium waiting to be colonized' (Lefebvre, 1974/1992, p. 125).

Space is not just what we see and feel materially. It consists of open plains and our imagination invested in these voids of space, and also its materialisation in human-made infrastructures. These infrastructures are in essence the navigational tools and architectonics of our everyday lives. They physically direct us in specific directions and limit us corporeally. One cannot walk through a wall or cross a border without showing a passport. Similarly, the social or cultural dimensions of space can limit or create opportunities. I cannot cross any border with just any passport, and even if I manage to cross the first human border controller with my passport, I might be stopped at the next electronic border when my face is recognised and correlated with a face in the crowd of people in a public demonstration.

The very design of an infrastructure is an active process. We do not just fill in a void of open space with an infrastructure – 'we infrastructure', as the information and Science and Technology Studies scholars Geoffrey C.

Bowker and Susan Leigh Star call it (Star, 1999; Bowker & Star, 2000; Star & Bowker, 2006). Practices of designing, repairing or even participating and navigating within infrastructures also imply actively participating in social power dynamics. Spatial infrastructures are the result of human controversy and negotiation; sometimes they even hold stories of violent and oppressive domination of one people or social group over another, as was, for example, the case of the infrastructural transformation of the North American prairies and the treatment of the indigenous Americans who originally lived and hunted on these plains.

As sites of social negotiation and power struggle, infrastructures are invested with different imaginations and hopes regarding the social appropriation of space (Larkin, 2013). This also means that the development of spatial infrastructures has often produced social conflict (Reeves, 2017). Langdon Winner (1980) famously used the low-hanging overpasses that connect Long Island to the rest of New York as a case in point. They were specifically designed to prevent access to public buses, which were used by the majority of black people, and thus, Long Island could be accessed mainly only by middle- or upper-class white people with private cars. In this way, one social group was prevented by infrastructural design from accessing the recreational areas of Long Island.

We can read infrastructures as 'narrative structures' of social power. An investigation of infrastructure should therefore, as Susan Leigh Star argues, always seek to restore these types of social narrative (Star, 1999, p. 377). Making the invisible factors of infrastructures visible by pulling the underlying (the 'infra') into the foreground also has a social function (Star, 1999; Bowker & Star, 2000; Bowker et al., 2010). It makes change possible and the hidden social consequences manageable (Bowker et al., 2010, p. 98). Nevertheless, as long as they run smoothly, these narratives of power are most often untold. A disruption, like the collapse of Morandi Bridge, however, will necessitate a more detailed explication of their inner working (Star, 1999).

To explore the infrastructural narrative of BDSTIs we can look at various incidents and moments where their integration in society was disrupted and their data power dynamics were exposed. In the early 21st century BDSTIs were intrinsically intertwined with organisational systems and practices (Ratner & Gad, 2019). They had evolved into sociotechnical infrastructures for the flows of global economies and societies, cutting across cultures and legal jurisdictions. BDSTIs were representing and constituting global societies and environments as the mundane background against which social practice, social networking, identity construction, economy, culture and politics were conducted. Yet in 2013, the Snowden revelations and the tirelessly journalistic efforts[1] that went into the subsequent investigation of the US National Security Agency's (NSA's) global mass surveillance system pulled the narrative of

BDSTIs into the foreground of public debate and exposed it in all its complexity. This revealed a material and immaterial global infrastructure enabling the mass collection of the personal data of European citizens by a foreign intelligence service. The leaked PowerPoint presentations from US intelligence officers detailing the PRISM program, published in *The Guardian* (2013), revealed how the mass surveillance intelligence system was intertwined with the largest global big data companies' social networking services. They also provided a detailed map of how the world's (Europe, US and Canada, Latin America and the Caribbean, Asia and Pacific, Africa) flows of data and communication were directed through nodes and hubs in the US. Phone calls, emails and chats would not take the most physically direct route, the PRISM slides showed, but rather the cheapest, which would go through the US. Data was collected through fibre cables and infrastructure as it flowed by, but also directly on the servers of the US-based companies servicing world users.

Following the Snowden revelations, the EU scrapped the legal framework for the data exchange infrastructure between the US and the EU market (the Safe Harbour agreement). They had revealed the core challenges posed by the global BDSTIs to traditional modes of territorial governance of world states and regions. In the years that followed the revelations, news of a range of other cases of infrastructural 'disruption' of the big data infrastructure emerged regularly from large data hacks and leaks from companies and institutions (for example, the Snapchat hack in 2013 or the Ashley Madison hack in 2015; see list of breaches on Wikipedia, 2020) to the revelation of the complex data analytics used to influence electoral votes (Cadwalladr, 2017; Rosenbergh et al., 2018, *The Guardian*, 2018). The incidents all caused momentary or long-lasting disruption to existing governmental and business BDSTI practices that had been imagined, conceived of and implemented since the 1990s. In particular, as I have argued elsewhere (Hasselbalch, 2019), due to this disruption and crisis of the BDSTIs, the big data imaginations and mindsets of the early digital developments were increasingly contested by alternative mindsets and imaginations about the conception and implementation of a European digital infrastructure.

The Narrative of a European Digital Infrastructure

In Europe, there is a space occupied by the imagination and interest of an EU project. It is a 'European infrastructure' that enables the efficient workings of a union of collaborating member states. The EU was historically created after the Second World War as an economic collaboration between countries, which was manifested in the idea of a single market. It later evolved into a political union around areas such as foreign policy, migration and security. According to this political project, a European infrastructure's architectonics should facilitate first and foremost a European union of member states; that is, European citizens' free movement as well as the European single market's free movement of goods and services. Thus, in European policy-making, 'infrastructure' is above all a term used to describe a system across Europe that enables cohesion and social and economic collaboration between member states. For example, the Trans-European Transport Network (TEN-T) policy was created with the objective 'to strengthen social, economic and territorial cohesion in the EU' (European Commission A, 2020) and concerns the implementation and development of a network of the EU's physical transport infrastructure, which in 2017 counted over 217,000 km of railways, 77,000 km of motorways, 42,000 km of inland waterways, 329 key seaports and 325 airports (European Commission B, 2020). The TEN-T network is part of a system of Trans-European Networks (TENs) that also covers energy and digital services. These programs are supported and implemented via the Connecting Europe Facility, which is a €30 billion (in 2019) funding instrument in the form of grants, procurement and financial instruments with the stated aim of 'further integration of the European Single Market' (European Commission C, 2019, p. 6). The European infrastructure is here defined in terms of its political purpose to support an imagined European community. To *do infrastructure* in the EU is a strategic political endeavour from which has emerged infrastructural practices – such as engineering and design standards, construction, investment and regulation – that produce space, or, in other words, that constitute 'engineered' and 'intended' components of a 'European infrastructure'.

In the 2010s, an effort to extend the material infrastructure of the EU with a digital sociotechnical infrastructure was increasingly voiced in EU official strategies and materialised in infrastructural practices, such as dedicated policies and investments. The EU's Digital Agenda intersected policy areas and regulatory frameworks that were traditionally treated separately (Valtysson, 2017). The term infrastructure was here still only used to describe the technical aspects of a digital infrastructure for the Single Market. However, although not described as such, the social and economic components of the European digital infrastructural architectonics gradually became a focal point of European policy and investment strategies.

In 2010, the Digital Agenda for Europe presented the 'Digital Single Market' as a new EU endeavour: 'It is time for a new single market to deliver the benefits of the digital era' (European Commission D, 2010, p. 7). The Digital Single Market aspiration was, among others, voiced as a response to a persistent fragmentation that was said to restrain Europe's competitiveness in a digital economy overshadowed by companies such as Google, eBay, Amazon and Facebook that 'originate outside the EU' (European Commission D, 2010, p. 7). A wide spectrum of infrastructural practices was therefore envisioned in the agenda to create and sustain the competitive space of the Digital Single

Market. Some were technical in nature, focused on 'interoperability' and the creation of technical standards, or the development of 'fast and ultra-fast internet access'. However, immaterial components were also described, such as ensuring the 'trust' of Europeans: 'the digital age is neither "big brother" nor "cyber wild west"' (European Commission D, 2010, p. 16).

Later, in 2015, a Digital Single Market Strategy for Europe was published to further enable the Single Market's 'free movement of goods, persons, services and capital' (European Commission E, 2015, p. 3). In his introduction to the strategy, the then-president of the European Commission, Jean-Claude Juncker, spelled out its foundational political purpose and imagining: 'I believe that we must make much better use of the great opportunities offered by digital technologies, which know no borders. To do so, we will need to have the courage to break down national silos in telecoms regulation, in copyright and data protection legislation, in the management of radio waves and in the application of competition law' (European Commission E, 2015, p. 2).

This was followed up with two communications published together in 2016 ('Digitising European Industry Reaping the full benefits of a Digital Single Market', European Commission F, 2016 and the 'European Cloud Initiative – Building a competitive data and knowledge economy in Europe') that each emphasised the impact and role of big data and spelled out the contours of a European BDSTI. In the first communication, big data was described as the foundation of an industrial revolution, and a new focus on a data-sharing cloud-based infrastructure for scientists and engineers in the EU also took form (European Commission, F, 2016 p. 2). It outlined concrete infrastructural practices to develop a data infrastructure in terms of investment, policy and coordination. In the second of the two 2016 communications, a concrete infrastructural initiative to support the development of a 'European Cloud' with an emphasis on a 'European Data Infrastructure' (European Commission, F, 2016, p. 8–10) was presented, and the BDSTI was specifically framed as a promise to strengthen the data and knowledge economy in Europe and use the potential of big data. It was first and foremost described as a data-sharing system consisting of various technical components: 'the data infrastructures which store and manage data; the high-bandwidth networks which transport data; and the ever more powerful computers which can be used to process the data' (European Commission F, 2016, p. 2). However, the European Cloud was not just a technical data infrastructure and organisation; it was also envisioned to facilitate the potential of big data by making 'it possible to move, share and re-use data seamlessly across global markets and borders, and among institutions and research disciplines' (European Commission F, 2016, p. 2).

Following these first depictions of a European data infrastructure, however, the aspiration to become a European Digital Single Market and data infrastructure competitor in a global market on similar terms as big data companies

gradually transformed into an aspiration to make this European data infrastructure and Digital Single Market a key differentiator in a global competitive
digital market. The aspirations to compete in a global big data economy while
preserving and protecting Europeans' fundamental rights were soon reconciled
in what was also referred to as the European 'third way', which I will describe
in more detail in Chapter 4.

Thus, in 2020 a general 'European data strategy' was proposed, with reference to a society 'empowered by data' and recognising the role of data in
society: 'Data will reshape the way we produce, consume and live. Benefits
will be felt in every single aspect of our lives, ranging from more conscious
energy consumption and product, material and food traceability, to healthier
lives and better health-care' (European Commission H, 2020, p. 2). Concrete
infrastructural practices were outlined, such as investment and policies
supporting the development of practitioner and user competences, European
science and research, technical data structures and data pooling, as well as the
development and possible implementation of legal frameworks to ensure the
development of the European BDSTI.

Data Ethics in European Public Policymaking

At a public debate in 2017 the European parliamentarian Sophia in't Veld said:
'I'm pretty convinced that the ethical dimension of data protection and privacy
protection is going to become a lot more important in the years to come' (in't
Veld, 2017). She was talking about the policy debates in Europe on digital data
and data protection regulation that were going on at the same time as policies
on the development of a 'European' BDSTI were gaining a foothold. You can
discuss legal data protection, she claimed, but then there is 'a kind of narrow
grey area where you have to make an ethical consideration and you say what
is more important' (ibid.).

In Europe, policy debates on the ethics and 'European values' of technological developments have been ongoing since the 1990s. The Council of Europe's
'Oviedo Convention' was, for example, motivated by what de Wachter (1997,
p. 14) describes as the 'feeling that the traditional values of Europe were
threatened by rapid and revolutionary developments in biology and medicine'.
The weight on *data* ethics, however, gained momentum in pan-European politics in the final years of the negotiation of the EU's General Data Protection
legal reform. As the European Data Protection Supervisor (EDPS) Digital
Ethics Advisory Group (2018, p. 5) described it in one report, their work was
being carried out against the backdrop of 'a growing interest in ethical issues,
both in the public and the private spheres and the imminent entry into force
of the general data protection regulation (GDPR) in May 2018'. What had

happened? How did this need to discuss and reflect on the ethical implications of data develop?

Throughout the early 2000s, the data protection field was transforming in the context of global information technology infrastructures, and new powers and interests in the data of these increasingly sociotechnical infrastructures were invested in the data being collected, transferred and processed. This was particularly evident during the years in which the GDPR was negotiated. Described as one of the most lobbied EU regulations (Warman, 8 February 2012), it was taking shape in the context of competing interests of economic entities, EU institutions, civil society organisations, businesses and third country national interests.

As I have described elsewhere (Hasselbalch, 2019), in the years following the first communication of the reform, data protection debates increasingly also included a reference to 'data ethics' in meeting agendas, debates in public policy settings, and reports and guidelines. After the adoption of the GDPR in 2016, the list of European member states or institutions with established data or digital ethics initiatives and objectives grew rapidly. Examples included the UK government's announcement of a £9 million Centre for Data Ethics and Innovation with the stated aim to 'advise government and regulators on the implications of new data-driven technologies, including AI' (UK Government, Digital Charter, 2018). The Danish government appointed a data ethics expert committee that I became a member of in March 2018 with a direct economic incentive to create data ethics recommendations to Danish industry and to turn responsible data sharing into a competitive advantage for the country (Danish Business Authority, 12 March 2018). Several member states' existing and newly established expert and advisory groups and committees began to include 'ethics' objectives in their work.

In an interview I conducted at the Internet Governance Forum (IGF) in 2017, a Dutch parliamentarian described how, in 2013, policymakers in her country were reacting to the impact of digital transformations on society, which they perceived as going 'very wrong' (interview, IGF 2017; Hasselbalch, 2019). She had already by then proposed the establishment of a national commission to consider the ethical implications of the digital society. As she told me, 'we need people to think about what to do about all of this, not in the sense, you know, "I don't want it", but more in the sense, are there boundaries? Do we have to set the limits to all of these possibilities that will occur in the coming years?'

In the official documents of many of the public policy data ethics initiatives launched after the General Data Protection Regulation came into force in Europe, data ethics was described as a response to emerging challenges of digital transformations and as a type of policy device to evaluate various policy options. These 'data ethics spaces of negotiation' (Hasselbalch, 2019) were created to make sense of challenges and conflicts surfacing when soci-otechnical changes challenged the state of affairs and established laws and norms did no longer provide a clear answer and solution to a problem. For example, a report from the EDPS from 2015 states: 'In today's digital environment, adherence to the law is not enough; we have to consider the ethical dimension of data processing' (p. 4). It continues by describing how various EU principles of law (such as data minimisation and the concepts of sensitive personal data and consent) are challenged by big data business models and methods. Depictions like these highlighted the uncertainties and questions that the new data infrastructures had created. European policymakers began to see digital data processes as meaningful components of social power dynamics. Information society policymaking thus was also increasingly perceived as an issue of the distribution of resources and of social and economic power, as the then EU Competition Commissioner Margrethe Vestager stated at a DataEthics.eu event on data as power in Copenhagen in 2016: 'I'm very glad to have the chance to talk with you about how we can deal with the power that data can give' (Vestager, 9 September 2016).

Power in the Information Society

In the early 21st century, the 'Information Society' was an established term and forceful strategic global policy focus. Most notably, this political agenda gained footing during the UN World Summit on the Information Society (WSIS). In 2003, the first summit was held in Tunis with the stated aim of gathering momentum and taking concrete political steps to establish the foundations for an inclusive Information Society and 'reflecting all the different interests at stake' (World Summit on the Information Society, 2013). Based on a sense of urgency and realisation that an ongoing digital revolution was transforming society as we know it, governments from around the world gathered at one of the UN's first multistakeholder fora to create a political agenda with the aim of tackling the societal, economic and cultural implications of the rapidly developing Information Society.

Something was disturbing the natural 'state of affairs' and there was a lingering feeling of a changing world, of symbolic and material borders disintegrating, and of necessary changes in modes of governance. The global policy environment saw this transformation of things and invested in it, and various authors have also depicted at length the particularities of a society based on 'information'. Frank Webster (2014) describes these preoccupations with the Information Society as a prioritisation of information. Despite the different views on what this prioritisation of information means, and what role information plays in societies, it is fundamentally, he argues, a way of conceiving something new and different about contemporary societies (Webster, 2014,

p. 8). He examines the literature that portrays the information society and finds five definitions, not necessarily mutually exclusive, but each emphasising different aspects of the new role that information is envisaged to play in society. Technological definitions are concerned with the evolution of specific 'new technologies' in society, such as the computer and IT technologies in general. Economic definitions focus on the economic value of informational activities, whereas occupational definitions track an increase in informational work. Spatial definitions examine the role of information networks in reshaping space and time, and cultural definitions of the Information Society look at the increasingly media-laden society and technological information environment (Webster, 2014, p. 10–23).

Here, I want to introduce a different perspective on the particularities of the Information Society. Rather than a prioritisation of information, we may consider all of these different definitions of what constitutes the Information Society horizontally in terms of a redistribution of power facilitated within new technologically mediated configurations of space and time, or what I refer to as BDSTIs and AISTIs. Thus, I explore the technological evolution of space for not just the exchange of information, but also as an architecture for global systems of power.

Initially, technological developments expanded our experience of space (Kern, 1983). With the telephone, one could, for example, be in two different places simultaneously. Later, with the introduction of the wireless, this simultaneity of experience was expanded to an instant sense of the whole world.

In the early 21st century – not many decades after Lefebvre's 1974 description of an occupied global space of commercial images, signs and objects – the digital evolution of a global geographical space was complete. The Google Maps service, for example, had transformed space into a digital data infrastructure with satellite imagery, aerial photography, street maps, and 360° panoramic views of streets combined with real-time traffic conditions and route planning.

Looking at the digital geographical evolution, of which Google maps is representative, sociology scholar Francesco Lapenta (2011) coined the term 'Geomedia' to describe emerging location-based services like Google Map's and Google Earth's merging of geographical space, virtual space and the local experiences of users based on big data and information exchanges. He describes these as mediating spaces that function as 'new organisational and regulatory systems' articulating and organising social interactions (Lapenta, 2011, p. 21). They are used by individuals as social navigation tools that can help reduce the complexity of global information systems to manageable and socially relevant information exchanges (Lapenta, 2011, p. 21). Geomedia is an example of the technologies that, in the 21st century Information Society, are transforming the bodily, social and individual experiences, physical space and location into interoperable digital data, blurring their lines of separation when integrated into the designed spatial architectures of a virtual infrastructure. As such, not only is our experience of space transformed but, as Lapenta argues, Geomedia also regulates social behaviour and interpersonal communications, as well as coordinating social interactions. As communications scholar Joshua Meyrowitz (1985) described in his momentous book *No Sense of Place*, our emerging electronic global and local information realities have real qualities that shape our social and physical realities. 'Information systems' modify our physical settings via new types of access to information, restructuring our social relations by transforming situations.

The Marxist cultural geographer David Harvey uses the term 'time-space compression' (Harvey, 1990) to describe the transformation of the human experience and thus representations of time and space in an increasingly globalised world. The annihilation of 'space through time' (Harvey, 1990, p. 241) is a reduction of distances between places in terms of travel time and costs. It is a shrinking world map representing a transformation of the very objective qualities of time and space. Harvey argues that space is conquered by humans through processes of 'producing space', of occupying and settling in space, and that these occupations of space are legitimised by specific legal systems that stipulate the different rights we have to the spaces we navigate in society. In this way, space also constitutes the internal and external spatial borders of a society. They are spaces occupied by human ideas that frame social processes and practices (Harvey, 1990, p. 258). Therefore, the transformation of time and space also has the function of maintaining power as it imposes the structure for social practices representing the forces of power in a given society. Harvey sees 'time–space compression' and the shrinking world map as not just consequences of technological developments per se, but also, importantly, as expressions of the embedded interests of the expansion of capitalism and industrialisation in the 19th century. Hence, while a 17th-century space was occupied with human ideas about a 'better society' and accordingly focused on a rational ordering of space and time to develop a society that guaranteed individual liberties and human welfare, time–space compression is, he argues, mainly created for the operations of capital and therefore designed for instantaneity, ephemerality, fragmentation, volatility and disposability (Harvey, 1990, 286–307).

We can use Harvey's idea of a 'space' that is open for 'active occupation' by different ideas and interests to consider the infrastructural practices that actively contribute to the material and immaterial shape, form, direction and orientation of power. This type of occupation works in a space of imagination and symbolic practices of power, and it works in very concrete terms as a property of the material space we design and create for ourselves in society. In this way, we can also say that in the early 21st century our space had a very material form of power, with global and local architectonics that had been imagined within specific business and political mindsets and ideas about the role, the opportunities and the challenges of information technologies.

The IT revolution was a major historical event, comparable to the 18th century's industrial revolution, argues another scholar of the information society, Manuel Castells (2010). He describes the 'Network Society' in terms of 'space of flows' (of capital, information and technology, organisational interaction, images, sounds and symbols) that work as the material form of power. It is a type of society in which dominant societal functions are organised around networks (Castells, 2010, p. 407–59) and this very architecture of flows constitutes the transformation of power (Castells, 2010, p. 445).

There are three layers to the 'space of flows'. Castells refers to the first layer as the 'material support' constituted by a 'circuit of electronic exchanges' in a global technological information network (Castells, 2010, p. 442), which also forms the foundation of an acceleration of the movement of people and goods. The second layer is constituted by 'nodes and hubs' (Castells, 2010, p. 443). The networks that enable the spaces of flows are not 'placeless', but rather organised around electronically linked-up 'places' with 'well-defined social, cultural, physical [and] functional characteristics' (Castells, 2010, p. 443). They have specific functions, such as exchange or communication 'hubs', or as the 'nodes' where strategically significant functions are located in positions with different constantly evolving hierarchies between them. The third material layer of the space of flows involves the spatial organisation and form created for dominant 'managerial elites' and takes its point of departure in the general notion that a society is 'asymmetrically organised around the dominant interests specific to each social structure' (Castells, 2010, p. 445).

The three-layered 'architecture' constitutes the foundation for an ongoing transformation of society, and it is in its very design that we may read 'the deeper tendencies of society, of those that could not be openly declared but yet were strong enough to be cast in stone, in concrete, in steel, in glass, and in the visual perception of the human beings who were to dwell, deal, or worship in such forms' (Castells, 2010, p. 448).

In Castells' depiction, societal power is concentrated in the very information architecture of technological networks. It is no longer fixed in places but distributed in the design of infrastructures of information flows: 'the power of flows takes precedence over the flows of power' (Castells, 2010, p. 500). Thus, to be connected or disconnected from the space of flows is the first step towards being empowered in the network society, while the second is to actively participate in the design and shaping of its global infrastructure.

2.   ETHICS IN THE SURVEILLANCE SOCIETY

Power transforms in the networks and flows that comprise the architecture of the Information Society. Politics and narratives about big data are invested in the very construction of the infrastructures that facilitate the particular shape of sociotechnical power of a Big Data Society.

BDSTIs are a form of power integrated in our spatial architectures. They are not liberating spaces in which human life flourishes. Most often they sustain the power of powerful actors in society while putting others at a disadvantage, and they are difficult to resist and change, particularly due to the very design of their data infrastructures, which track and monitor personal data by default and restrict citizens' liberty and agency. In other words, the acceleration and integration of what I have referred to elsewhere as 'Destiny Machines' (Hasselbalch, 2015) in ordinary state and business practices has resulted in a complex and advanced societal machinery of data power that leads, guides and defines human lives. A data ethics of power addresses the ethical challenges of these particular structures of power.

Destiny Machines are technological systems and processes designed to predict human behaviour based on the accumulation of personal data and then act on these predictions. Every day humans interact with these big data 'machines', designed to predict human behaviour by tracking, scrutinising and analysing a recorded and stored data past and present of the 'data doubles' of humans (Haggerty & Ericson, 2000). In this way, human lives are framed and pointed in specific directions. Destiny Machines can also be described as machines specifically designed to produce machine-readable people and shell out destinies on the other side of the production line. In fact, they produce, create, act on and define destinies. We might even say that fate is what Destiny Machines produce. This is what is being innovated with; it is part of an actual machinery and can literally be sold and traded (Hasselbalch, 2015). Within this machinery, human lives are made 'programmable' (Frischmann and Selinger, 2018) and meaningful only within the sorting structures of inclusion or exclusion of the surveillance assemblage (Lyon, 2010).

In fact, the Destiny Machine does not need the human life and body; it is fundamentally indifferent to the individual human being as only our 'data doubles' are meaningful within its surveillance assemblage. Or, said another way, it only has an interest in the 'data derivatives' (Amoore, 2011) of the 'data double'. As professor of political geography Louise Amoore describes, it is 'not centred on who we are, nor even on what our data say about us, but on what can be imagined and inferred about who we might be – on our very proclivities and potentialities' (Amoore, 2011, p. 24).

It is what the sociologist David Lyon refers to as the 'surveillance society' (Lyon, 2001, 1994), or more specifically a 'liquid surveillance society' (Lyon, 2010; Bauman & Lyon, 2013) sustained by sociotechnical 'data flows' (Lyon, 2010, p. 325). 'Liquid surveillance' has a different shape than the form of surveillance that Jeremy Bentham famously outlined in his description of the panopticon prison (1787) and Michel Foucault's panopticism (Foucault, 1975/2018), which is centrally integrated in the spatial architecture of society and enforced as a type of aware self-discipline. It does not come from a centralised visible above ('sur'), or middle (Bauman & Lyon, 2013), but is embedded in digital infrastructures, networked, distributed and sustained by increasingly greater distances between those that watch and those being watched (Galic et al., 2017). Opaque and bottom up, liquid surveillance is invisibly intertwined with individuals' lives, and therefore, it is also inscrutable and generally difficult to address (Lyon, 2010). Notably, surveillance is not exceptional, but a condition of experience and human life in the surveillance society. It is a culture (Lyon, 2018) based on 'dataveillance', a systematic monitoring, tracking and analysis of personal data systems (Bauman & Lyon, 2013; Clarke, 2018; Christl & Spiekerman, 2016). Taking form as an 'assemblage', it abstracts the human body from a digital 'data double' which can be scrutinised and used for purposes of control by governments or can be sold for profit in commercial interchanges (Haggerty & Ericson, 2000).

Transformation of the actors of power in the surveillance society also means widening the attention of ethical scrutiny from a focus on the more conventional arbitrary surveillance powers of states to the commercial stakeholders that gain power through accumulation, tracking and access to big data. Surveillance is a 'surveillance-industrial complex' in which it is the very sociotechnical interrelation between state and private sector actors that makes surveillance possible (Hayes, 2012). Our private lives become part of a public–private space prone to intelligence gathering activities which are often legally possible, but ethically problematic (Røn & Søe, 2019). This change in power dynamics is a core ethical problem, as it is based on an increasing information asymmetry between individuals/citizens/workers and the powers of the big data companies that collect and process data in digital networks (Pasquale, 2015; Powles, 2015–2018; Hasselbalch & Tranberg, 2016; Zuboff, 9 September 2014, 2019, 5 March 2016; Ciccarelli, 2021). As Tranberg and I illustrated in our book, 'the biggest risk lies in the unequal balance of power that the opaque data market creates between individuals and corporations' (Hasselbalch & Tranberg, 2016, p. 161).

Harvard professor Shoshana Zuboff (2019) describes 'surveillance capitalism' as an accumulation of a capitalist logic that commodifies human psychology and experience to satisfy market forces and the commercial aims of tech giants. Her work on surveillance capitalism raised public attention in the late 2010s to the role of named powerful Silicon Valley industrial actors, such as Google and Facebook. Her main concern is the way in which these new commercial forms of digital surveillance reshape the institutional structure of modern democracies, and she describes this form of power in very concrete terms:

> Two men at Google who do not enjoy the legitimacy of the vote, democratic oversight, or the demands of shareholder governance exercise control over the organization and presentation of the world's information. One man at Facebook who does not enjoy the legitimacy of the vote, democratic oversight, or the demands of shareholder governance exercises control over an increasingly universal means of social connection along with the information concealed in its networks. (Zuboff, 2019, p. 127)

Big Data Ethics

An urgent call for data ethical action on the power of big data and algorithms takes form in critical studies of contemporary sociotechnical structures of surveillance and power. Lyon emphasises the urgency of developing an 'ethics of surveillance' (Lyon, 2010, p. 333), or what he, in conversation with the sociologist and philosopher Zygmunt Bauman, also refers to as 'surveillance ethics', to address the 'political realities of surveillance' (Bauman & Lyon, 2013, p. 20). They identify two major issues addressed by such an ethics: one they refer to with Bauman's term 'adiaphorization' (Bauman, 1995), where morality is abstracted from the very systems and processes of surveillance, while the other is the distance created between the human being and the consequences of their actions (Bauman & Lyon, 2013, p. 7). It is a very practical applied data ethics, an 'ethics of Big Data practices' (Lyon, 2014, p. 10) aimed at renegotiating what is increasingly exposed to be an unequal distribution of power between individuals and the institutions that develop BDSTIs. As Lyon later states, with direct reference to the 2013 Snowden revelations:

> We need ethical tools for assessing surveillance, a broadened sense of why privacy matters and ways of translating these into political goals. And it is essential that we do this with a clear sense of what kind of world we are working towards. How do we get a sense of what a better world would look like? (Lyon, 2014)

Studies of the surveillance architecture of the surveillance society purposely seek to disclose new constellations of power in sociotechnical assemblages and to hold those in power accountable. Eric Stoddart argues that surveillance studies, as a discourse of disclosure, constitutes a method of ethical enquiry (Stoddart, 2012, p. 369). Here he considers two strands of approach that ethically evaluate surveillance. The first is a 'discursive-disclosive' approach that seeks 'to disclose what is being done and the possibilities that might be available for alternative actions' (Stoddart, 2012, p. 372). He refers to a type of Foucauldian ethics in which ethical inquiries address practices of surveillance, rather than only processes and, as such, he emphasises ethics as a process of liberating reflection. As he describes it, 'A discursive approach discloses to both us and others what we did not previously know about our situations, the conditions under which we have been living and working and how we might be being exploited' (Stoddart, 2012, p. 372). The second approach is what he refers to as a 'rights-based' approach, with reference to a body of human rights work that demands the 'accountability of those with the power to watch' (Stoddart, 2012, p. 369). We can here think of the new power actors of a 'platform economy' shaped by the large digital technology platforms of the 21st century that transform the structure for – and accordingly the adequacy of – traditional modes of the protection of individual rights (Belli & Zingales, 2017; Wagner et al., 2019; Franklin, 2019; Jørgensen, 2019).

Along these lines, legal scholars Neil M. Richards and Jonathan King (2014) suggest a more inclusive rights-based analysis based on a 'big data ethics' (Richards & King, 2014, p. 393) that points to the ethical implications of the empowerment of institutions that possess big data capabilities at the expense of 'individual identity' (Richards & King, 2014, p. 395). In this way, when addressing the distributed power relations of the Big Data Society as a condition for the implementation of the right to privacy, for example, we may also better understand privacy as 'contextual' (Nissenbaum, 2010), effected and created in groups (Mittelstadt, 2017), and therefore a collective, rather than only an individual, responsibility (Tisne, 2020).

I draw the very material of a data ethics of power from these depictions of the surveillance properties of a sociotechnical data-infused environment. The spatial architecture of BDSTIs, and also AISTIs, is by design sustaining asymmetries of power, but they are not just reinforcing existing power dynamics – they are also creating new structures and actors of power. Commercial actors gain power with the accumulation of data and data design made for purposes of profit and/or control challenging traditional state power.

A data ethics of power addresses this new natural state of power in the Big Data Society and calls for an alternative design and implementation of data systems; but more importantly it desires different data cultures (a term I return to Chapter 4). Here, 'surveillance capitalism' (Zuboff, 2019) incapsulates the role and power structures of capital and commercial actors very well. Yet, this idea does not capture the 'liquidity' (Bauman, 2000; Bauman & Haugard, 2008; Castells, 2010; Lyon, 2010; Bauman & Lyon, 2013) of the power structures and cultures of a Big Data Society, which is what a data ethics of power needs to address. That is, a power that is indeed concentrated and engineered by a few power actors, yet also increasingly self-sustained, re-engineered and evolving in (surveillance) cultures (Lyon, 2018) of use, design, governance and imagination, and therefore difficult – but not impossible – to change. I propose that it is exactly this 'liquidity' of power that necessitates a holistic data ethical governance approach to the Big Data Society.

The Asymmetric Experience of Data Power

During the COVID-19 pandemic, online global data maps, such as the John Hopkins University COVID-19 Dashboard, monitored and categorised regions in red zones of deaths and disease. Concurrently, the heterogeneous patterns of the erratic movements of the deadly virus materialised in experiences of segregation and exclusion of 'red zone' populations or communities with suddenly imposed social controls, and even discrimination and abuse (Xu et al., 2021).

In the 21st century, all lives are part of a global surveillance assemblage and consequently the experience of risk of sudden subjugation to power is also a shared human experience. The ephemeral and volatile nature of the digital means that we are all exposed and all at risk. Yet, the direct experience of social structural data power as a constant and certainty is not new and far from homogeneous (Lyon, 2007; Browne, 2015).

'Surveillance is nothing new to black folks', as Simone Browne (2015, p. 10) puts it, when describing the experience of surveillance of African Americans, who throughout history have been subjugated to acts of surveillance, violent branding and control:

> [R]ather than seeing surveillance as something inaugurated by new technologies, such as automated facial recognition or unmanned autonomous vehicles (or drones), to see it as ongoing is to insist that we factor in how racism and antiblackness undergird and sustain the intersecting surveillances of our present order. (Browne, 2015, p. 8)

Existing historically rooted inequalities are replicated in data systems of power. Powerful tools are used by powerful actors, and already exposed and vulnerable communities continue to be the objects of systems of arbitrary power. They do not become the 'managerial elite' (Castells, 2010). They might be hired in the machinery as low-paid content moderators under harsh working conditions (Chen, 23 October 2014) and to provide their data resources and attention (Advocates for Accessibility, 3 May 2020). Yet rarely are they the subjects of digital empowerment. And, as I will also argue at the end of this chapter, seldom do they have a voice when solutions and governance are proposed and developed.

On a global scale, countries and regions are divided between the 'information rich' and 'information poor'. Many developing countries did not experience the digital information revolution of their economies and societies. Instead, their populations gained access to the scraped version of the internet, FreeBasics, offered by Facebook in exchange for the intensive collection of their data (Advocates for Accessibility, 3 May 2020). Once again, being 'rich' or 'poor' will generally determine your level of exposure to and experience of arbitrary powers. As the political scientist Virginia Eubanks has illustrated, automated decision-making in social welfare provision in the US is a sophisticated evolution of the 19th-century poor houses (Eubanks, 2018). Examples of data power and the discriminatory treatment of people with socially challenging demographics are to be found everywhere. In 2020 in the UK, an algorithm that weighted in a school's historical performance when grading individual students caused the grades of students from large state schools to plummet, while the grades of students from smaller fee-paying private schools increased (Hern, 21 August 2020). In the Netherlands it turned out that the fraud detection system, SyRi, used by government agencies was predominantly applied in low-income neighbourhoods (AlgorithmWatch, 2020).

Today we all have a digital 'data double' (Haggerty & Ericson, 2000) abstracted from the human body that can potentially be used for purposes of scrutiny and control by governments, sold for profit in commercial interchanges and also targeted by direct acts of digital violence and abuse. However, the direct experience of 'data violence' (Bartoletti, 2020) is in practice generally not shared. A Danish woman, Emma Holten, described her experience of having intimate images shared online without her consent and the brutal response from countless men in the years following as a vicious form of 'objectification' (Holten, 1 September 2014). Online revenge porn and aggressive exposure of one's intimate life is predominantly a female fear and experience. Other direct experiences of data oppression are bound to one's ethnic heritage; the experience of disempowerment of Black people represented online as 'dead', 'dying' or 'detained' (Noble, 2018), 'sexual' (Noble, 2018), 'criminal' (Sweeney, 2013); or similarly people with specific ethnic biomarkers defined as less beautiful (Levin, 8 September 2016). These are everyday experiences to some people.

While surveillance technologies per se are frowned upon in a Western context as a challenge to the exercise of democratic citizen rights, there are always accepted exceptions that allow their design and application in order to manage particular risks and solve social problems. When a society defines a specific group or community of people as a 'problem' to be solved, for example, data technology and systems will be designed to target this problem in as sophisticated a manner as we allow it to be. Biometric identification technologies and systems are no exception to this rule. In Europe, entire communities of 'third country nationals' or 'stateless' people ('refugees' and 'migrants') have been identified as constituting a European 'migrant crisis' and an extensive migration and border management system has been put in place. Thus, for example, when asylum seekers or migrants cross a border, their fingerprints are collected in a specifically designed EURODAC database for 'unauthorised' accesses to a country. In addition, in the case of visa and citizenship applications or migration procedures, including asylum procedures, a centralised system – the ECRIS-TCN system – for the exchange of criminal conviction data on third country nationals and stateless persons is in place. The ECRIS-TCN will, for example, allow for the processing of fingerprint data for the purpose of identification. At the time of writing the expectation was that facial images would also become part of the EURODAC database and the ECRIS-TCN system for identification purposes, with the use of facial recognition technology (Wahl, 10 September 2019). A regulatory proposal in 2020 even proposed the lowering of the age for biometric data collection on children in EURODAC from 14 years of age to six years (Marcu, 29 April 2021). While data power is increasingly externalised, disinterested and sophisticated, the very power dynamics have not changed. We might all live in liquid forms of surveillance, but we do not all experience the reality of it.

Does Data Ethics Have a Voice?

In 2020 the Netflix documentary, *The Social Dilemma*, cast light on the ethical implications of big technology companies in our everyday lives, reaching a new audience on one of the world's most popular online streaming services. Although an important critique, the problems identified appeared strangely new and surprising in the film. It was a story told by only North American characters, and with a predominantly white male voice in a 'damage control' mode. The silence of the people most exposed to data power was glaring. The silence of a global civil society movement was equally so. The film did not feature the key agents of change from regions and cultures all over the world that had worked for decades on the disclosure, the alternatives, the law and the awareness of the arbitrary global data powers of the big tech companies.

A data ethics of power is not only about power, it *is* power. The power to raise issues, to define problems, to propose and create the solutions to the problems. The empowerment to reject the objectification of the digital surveillance assemblage and raise one's voice against it as a subject. Yet rarely do we hear the voices of the people who are subjugated to data power when solutions are proposed (Levin, 29 March 2019). The engineer in command of the 'solution' to the 'ethical problem' in the engine room in *The Social Dilemma* is the white male actor, Vincent Kartheiser, in a white suit with a fluent North American accent.

When examining some of the personal accounts of people with particularly brutal experiences of data power, I found that core to the experience of abuse was a feeling of disempowerment, the inability to speak and act against power. As Emma Holten says in her description of her experience with the spreading of her intimate images online:

> To use women as sex objects for one's own pleasure, **without them having anything to say about it**, is everyday life online. (my translation and emphasis) (Holten, 1 September 2014).

The UK student, Laura Hodgson, who received lower grades than she expected and deserved because of an algorithm that weighted in the historical performance of her state school, formulated her sense of disempowerment in a letter to her government accordingly:

> I write to you as a A-level student who has just received their results. I am devastated and upset by the results I have been given as the result of a system that was forced upon me. **I had no say in any of this**, yet I am expected to just live with these unfair grades. (My emphasis, Gill, 13 August 2020)

Not only does the inability to speak up constitute a feeling of disempowerment, it is also often associated with fear of escalation of power and abuse, as it comes with a cost to speak up from a vulnerable position. As the black American man, Robert Williams, who was arrested by the police due to an incorrect match between his face and another black man's face by a facial recognition system, writes in an op ed:

> As any other person would be, I was angry that this was happening to me. As any other black man would be, **I had to consider what could happen if I asked too many questions or displayed my anger openly** – even though I knew I had done nothing wrong. (My emphasis, Williams, 24 June 2020)

This is evident in politics as well. Being a highly visible female politician also means being more frequently targeted by 'uncivil messages' online (Rheault et al., 2019).

Thus, a data ethics of power does not only have a voice, it is also a silence that can be heard only if we choose to look for it and include it in spaces of negotiation. And if we do listen, we will find that these experienced voices come with incredibly valuable and nuanced solutions and approaches. Three years after her experience with online abuse, Emma Holten decided to reappropriate her body and created what she calls 'a new story about my body'. With a new set of photographs that she shared online she gave her naked female body a personal voice as a subject in control (Bødker, 1 September 2014). At that time, risks to online privacy were identified in the European internet governance debate mainly in the abstract, as a threat to markets and democracy, and solutions were carried forward loudly and even at times aggressively as questions of mostly technical design and legal requirements. Emma Holten's voice came into this debate as an alternative voice leaving a powerful mark based on her experience of abuse and discrimination as a woman online. My point is here that we see nuances, we restate our problems and we find new solutions when social and personal experiences are voiced. For example, the technical is essential in a discourse on the way in which privacy is challenged in the context of online sociotechnical infrastructures. It will consider 'stack-issues', 'application-requirements', 'decentralised' technical infrastructures, 'asymmetric upgrade' and 'privacy biometrics' as solutions to macro-power imbalances in democracies and markets. However, the 'technical debate' often seems to be disinterested in the very cultural experiences of abuse and thus in the nuances of the complex sociotechnical processes necessary to effect change.

The change necessary to counter the violent and oppressive effects of intensive data collection and processing in terms of loss of opportunities, discrimination and skewed cultural representation comes with the lived experience of being targeted and outside. It is a voice that is most often not the loudest one. It is a voice that knows abuse but does not necessarily understand the very technicalities of the abuse – the data design of it, the data analytics behind it – because it did not have a voice in its design. It is a voice that is regaining power by reappropriating the debate and the abuse on its own terms. Often it does not speak in a language that the West understands, or it comes with a broken accent, and with a plethora of experiences from multiple regions worldwide. Importantly, it is a voice that does not accept without question highly technical solutions to the complex problems and implications of big data and AI systems. This is because data abuse is not a technicality. Data power and abuse are social, cultural, historical and not the least personal experiences.

In the next chapter I will look at 'data ethical governance'. Core to this concept is the constitution of a critical space of value negotiation characterised by the inclusion of multiple actors in governance processes. Importantly, the foundation of data ethical governance is an awareness of the very conditions of power between different stakeholders. Thus, inclusion of stakeholders is not a straightforward process. For example, I used to work for years with the inclusion and empowerment of young people in the internet governance public policy debate where I learned that just adding young people in a policy panel debate rarely made much of a difference. On the other hand, in the workshops and the focus group surveys we did with young people, they could speak openly and freely among their peers, and we would receive valuable input for the internet governance policy processes we were engaged in (see, for example, Hasselbalch & Jørgensen, 2015). The conditions we create in governance spaces of negotiation are core to meaningful participation and inclusion. We need to expect and accept that people with different experiences will present their ideas in untraditional ways that might not be pleasing to an accustomed ear and that might not make sense in traditional governance contexts. However, they will have real stories that will challenge our ways of stating a problem and finding a solution. We will have to accept that alternative views and discourses are good for governance as they will help reshape the very conditions in which problems are stated. There are countless ways of engaging and ensuring that a plurality of voices are heard and accepted in processes of governing sociotechnical change. Very often, it is not only a question of simply composing representative groups and initiatives: it is also about empowering civil society and minority communities with the resources to participate and the competences to compete with the most powerful interests shaping the institutional politics of AISTIs and BDSTIs.

NOTE

1.   By journalists such as Laura Poitras, Glenn Greenwald, Henrik Moltke and others.

2. Sociotechnical change and data ethical governance

'Morality and law, on the other hand, begin when controversy arises. We invent both when we can no longer just do what comes naturally, when routine is no longer good enough, or when habit and custom no longer suffice.'
Richard Rorty, 1999

The philosopher Richard Rorty (1999) describes morality as a response to a disturbance or change of relations. When our routines and habits are challenged, and when we question the social construction of what we once knew, only then 'we shall confine ourselves to debating the utility of alternative constructs' (Rorty, 1999, p. 86). That is, ethical reflections do not come from a transcendental ideal; they emerge out of their relations with other things, things that are transformed and juxtaposed.

In this chapter we look at the role of data ethics in governance and sociotechnical change. In our daily lives, sociotechnical infrastructures are mundane and we take them for granted. They have no visible being as spaces of moral and ethical compromises. Yet, when they break down, malfunction or clash with other legal or moral systems, their embodied moral compromises become visible. This moment when the narrative of one system becomes visible, as we saw in the previous chapter, is also the moment of negotiation that will direct the development of a new system and the transformation of the old. We can identify moments on the macro-scale of time of a developing sociotechnical system that are prone to the 'ethical governance' of sociotechnical change. These moments, as I argue in this chapter, emerge in between crisis and consolidation in society (Hughes, 1983; Moor, 1985). They are critical moments, as they constitute social negotiation and result in the cultural compromises or the 'technological momentum' (Hughes, 1983, 1987) that a sociotechnical system needs to evolve. They are also crucial to phases of innovation and development, as they constitute the transformation of the sociotechnical system that emerges out of a quest to solve critical problems of that system.

In the 2010s, BDSTIs and AISTIs were developed and adopted rapidly in the public and private sectors, and their social and ethical implications became apparent at the same speed. As a result, the very design of these sociotechnical systems was increasingly questioned in policymaking and public debate, and new technical designs and business models, as well as legal and social requirements, emerged. This process also included the development of a new normality for information computer scientists' work on BDSTIs and AISTIs, with a new institutionalised framework consisting of standards and laws that directly addressed the ethical and social implications of their work with data. We were in an 'in between' moment in which the critical sociotechnical problems of the existing technological cultures of BDSTIs and AISTIs became visible and interests and values were explicated and negotiated in 'data ethics spaces of negotiation'. Consequently, in Europe specifically, notions of 'ethical' and 'trustworthy' technology were integrated into the politics, innovation and development of a new system.

1.   SOCIOTECHNICAL CHANGE

How do technological systems change? What constitutes sociotechnical transformation? Sociotechnical change is not just an arbitrary evolution of a system. It consists of a variety of components that can all be subjected to human shaping, direction and governance. In this book we want to know more about the way in which a human ethics can influence sociotechnical developments, but in order to effect such a change, we need a conceptual basis for the complexity of factors that constitute sociotechnical change.

As a point of departure, technology is an expression of social practice, created in a dynamic relation between human and nonhuman actors and factors in technological, material, social, economic, political and cultural environments (Hughes, 1983, 1987; Bijker et al., 1987; Misa, 1988, 1992, 2009; Bijker & Law, 1992; Edwards, 2002; Harvey et al., 2017).

The internet, for example, constitutes a type of applied science and knowledge written down in models, manuals and standards and practiced by engineers and coders, but it is also the result of use in various cultural settings, and it embodies social and legal requirements, political and economic agendas, as well as cultural values and worldviews. That is, technological or scientific processes are not objectively given or representative of natural facts and a natural state of affairs. On the contrary, these very processes and facts are conditioned by their place in history, culture and society and may also be challenged as such.

Paradigm Shifts in Scientific Knowledge and Practice

One way of understanding change is to examine paradigm shifts in the foundational knowledge, scientific and engineering practices that are invested in the design and development of technology. Thomas Kuhn (1970) famously examined the historical factors that form paradigm shifts in science, or what he described as 'scientific revolutions', that shatter the tradition-bound practices

of what is considered 'normal science' (Kuhn, 1970, p. 6). He argued that
when big shifts in a scientific field occur, it is not just a question of one theory
disproving another, or because a major scientific advancement has been made.
Scientific paradigm shifts also involve different ways of seeing the world and
doing science accordingly (Kuhn, 1970, p. 16). Scientific revolutions therefore
implicate shifts in the very foundational knowledge paradigms of scientific
practice. Kuhn described scientific practice in terms of particular conceptual,
observational and instrumental applications of knowledge that can be traced in
a specific scientific community (Kuhn, 1970, p. 43). Put in another way, a paradigm in science represents a way of doing science in a particular scientific
community according to a particular social and cultural worldview and priority
setting. This means that when a 'scientific revolution' occurs in a field it is not
only the science that changes. Fundamental change happens in regard to what
a scientific community counts as a problem worth solving, the 'scientific imagination' of the field and the kind of educational and instrumental environment
that the science is applied in. Accordingly, scientific paradigm shifts are major
and revolutionary in that they uproot everything that is considered normal
in the field, creating not just new theories but also new practices, standards
and methodologies, instruments and objectives (Kuhn, 1970, p. 6) They are
'earth-shattering', transforming normality by changing the very foundations of
what we assume to be a problem, a purpose, a shape and direction. As Albert
Einstein described the early moment of the scientific revolution of physics
he spurred: 'It was as if the ground had been pulled out from under one, with
no firm foundation to be seen anywhere, upon which one could have built'
(quoted in Kuhn, 1970, p. 83). These kinds of changes encompass not only the
acceptance of a new normal in how a scientific field is practiced and conceived
of, but also in how it is governed in society, and accordingly it requires the
development of new scientific practices, methodologies and standards.

The Complexity of Change

Considering technological change in terms of scientific paradigm shifts is
helpful when examining how the institutionalised standards, as well as the
foundational worldviews and knowledge frameworks of the people who build
technological systems, transform and how technological practices change.
However, an analysis that focuses only on the transformation of technological
components as forms of applied science would be incomplete. A diverse and
complex set of social, political, economic, cultural and technological factors
come together when large sociotechnical infrastructures, such as BDSTIs
and AISTIs, transform and evolve in society. Understanding technological
change means discerning the complexity of these components in terms of their

negotiations, and thus, the compromises that they embody. As the science and technology scholars Wiebe E. Bijker and John Law once stated:

> Technologies always embody compromise. Politics, economics, theories of the strength of materials, notions about what is beautiful or worthwhile, professional preferences, prejudices and skills, design tools, available raw materials, theories about the behaviour of the natural environment – all of these are thrown into the melting pot whenever an artefact is designed or built. (Bijker & Law, 1992, p. 3)

Understanding the diversity of factors that constitute the shape of a sociotechnical system also empowers human actors to direct its evolution. One way of doing this is to explore the different interests invested in sociotechnical change as well as their conflicts and negotiation. This also implies an examination of the moments of controversy and conflict in which core problems of technological systems are identified; solutions to problems, successes or failures of systems are negotiated; and priorities and goals for the evolution of the sociotechnical system are set (Hughes, 1983; Hughes, 1987; Misa, 1992). Following these controversial moments, a sociotechnical system is stabilized, generally accepted, and consolidates in society. It becomes, so to speak, the state of affairs. It is this focus on the very conditions for technological change that makes the trajectory of a technology's development and societal adoption manageable by humans. As Francesco Lapenta (2017) states, the future is 'not arbitrary but the product of a complex series of decisions and actors that can potentially give shape to a number of differently possible, probable, or desirable future scenarios' (Lapenta, 2017, p. 154). That is, the often conflictual negotiations about the technological development of the present (such as the ones we saw, for example, in the public debates on BDSTIs and AISTIs in the 2010s) must also be thought of as reflective choices about the future. Here, I want to argue that reflection on the ethical compromises and the trade-offs between different values and interests we make should be a core component of these moments of controversy.

The Four Phases of Sociotechnical Change

The transformation of sociotechnical systems can be studied in the way in which their different components evolve in patterns over time. Grasping developments on a macro scale allows us to intervene in moments open to critical intervention to shape the direction of evolving sociotechnical systems.

A key theory of the change of large sociotechnical systems is Thomas P. Hughes' (1983) analysis of the phases of the development and expansion of the world's electric power systems between 1880 and 1930. With a description of the specific complex economic, political, social and scientific components of the different phases of this development, he also illustrated more generally how technological systems evolve in patterns over time in constant dialogue with the interests embedded in their environments.

Hughes argued that although larger sociotechnical systems are instituted in different places and reach the different phases of development at different times, they evolve and expand according to a model pattern consisting of phases that are characterised by their dominant activities: invention; development and innovation; transfer; growth; competition and then, finally, consolidation (Hughes, 1983, 1987).

> **Invention, development and innovation:** The first phase is characterised by inventors and entrepreneurs that are the key drivers for the invention and the initial development of the system.
>
> **Transfer:** In the second phase the focus moves to the process of transferring the technology from one region and society to another and equally the dominant agents of change involved in this phase include, in addition to the entrepreneurs and inventors, the financiers and organisers of enterprises as key actors.
>
> **Growth:** In the third phase a range of actors, entrepreneurs, inventors, engineers and others, dedicate their efforts on correcting and finding solutions to 'reverse salients' that are formulated as critical problems that prevent the system from growing.
>
> **Momentum, competition and consolidation:** A large sociotechnical system requires a momentum with 'mass, velocity and direction' which is created by the different interests invested in the system in the fourth phase of societal consolidation.
>
> (Hughes, 1983, p. 14–15)

Particularly the third and fourth phases of the development of larger technological systems are helpful when describing the development of BDSTIs and AISTIs in the late 2010s. To start with, Hughes refers to 'a battle of the systems' in which an old and a new system exist simultaneously in a relationship of 'dialectical tension' (Hughes, 1983, p. 79). He describes this third phase as a moment of conflict and resolution, not only among engineers but also in politics and law (Hughes, 1983, p. 107). In these moments of conflict, critical problems are exposed, different interests are negotiated, and then they are finally gathered around solutions to direct the evolution of the systems.

In the growth phase just before consolidation, 'reverse salients' are formulated as critical problems. A 'reverse salient' is a component of an expanding system that 'does not march along harmoniously with other components. As the system evolves toward a goal, some components fall behind or out of line' (Hughes, 1983, p. 79). Accordingly, in this phase there is also an intense focus on problem identification, and solutions are proposed and negotiated by various actors. The new system, or the transformation of the old system, evolves out of the very problems identified and solved in this phase. A 'reverse salient' may be technical problems, but they may also be financial or organisational, and once identified, a group of 'problem solvers', from inventors, engineers and managers to financiers and legal experts, takes over to create solutions for them (Hughes, 1987, p. 74). 'Reverse salients' may arise from inside the technical systems or from their immediate environment, but crucially they are bound by time and place (Hughes, 1983, p. 80). In other words, the critical problems of the systems are not just resolved as technical problems, for example, with agreement on technical standards with systems requirements, but they are in dialogue with political and historical factors. In contrast to Kuhn, Hughes does not describe the phase of conflict and resolution as necessarily a revolutionary one. Change does not take form as the replacement of one paradigm with another that is incompatible with the first. The systems change in 'synthesis' and in a combination of 'coupling and merging' between the old and the new systems, which gradually evolve over decades and on different levels from the technical to the institutional, with invested interests gradually transferring from one system to the other (Hughes, 1983, p. 121). Only in cases where a 'reverse salient' cannot be resolved within the system does the solution need to be found in the development of a radically new system (Hughes, 1987, p. 75).

Now, applying Hughes' description of particularly the third and the fourth phases of sociotechnical change to the development of BDSTIs and AISTIs in the late 2010s, there are recognizably similar patterns. In the early 2010s, a global big data digital infrastructure connected different regions of the world, cutting across jurisdictions and thus challenging their traditionally territorial scope (see Hasselbalch, 2010). Most profoundly, the legal rights to privacy and personal data protection were challenged by this new technologically enabled interjurisdictional space in which different levels of protection and safeguards were required and implemented in the design of data technologies and systems. Clashes between different regions' legal frameworks for protecting privacy and for protecting business or state interests in data emerged, and various privacy-by-design technologies and systems were initially proposed as solutions to the critical problems of the 'old system'.

In the mid-2010s, critical problems concerning privacy rights and personal data protection became, as described in the previous chapter, particularly prevalent following Edward Snowden's revelations of mass surveillance and major data hacks of online services, such as the social networking service Snapchat in 2013 or the infidelity site Ashley Madison in 2015. Such critical problems were revealed and identified by activists, whistleblowers and journalists and picked up by engineers and policymakers from different regions of the world, who would propose, impose and design technical and legal solutions (such as Caspar Bowden, Max Schrems and many others).

Here, one can consider Hughes' depiction of 'reverse salients' as components of a system that fall out of line or are disharmonious with other components of the system, and therefore freeze the system's consolidation in society (Hughes, 1983, p. 79). The critical problems of BDSTIs, and also of AISTIs, that surfaced at this time, particularly regarding the protection of personal data and privacy, did indeed halt the ongoing consolidation of BDSTIs, causing first and foremost a battle between different regional legal governance approaches to the technical development and business conduct behind these systems. These were the 'reverse salients' of BDSTIs restricting a global big data system in growth and consolidation and, importantly, limiting its momentum in society with conflicts between business interests, citizen interests, state agency interests, and political and regional interests.

For example, a series of big data social networking services developed predominantly in one US area, namely Silicon Valley, had in the 2000s cut across the globe practically unnoticed by legislators, and therefore in the 2010s were already silently consolidating in European people's social and private lives (for example, 44% of Europeans said they never used social networking services in 2011 [Eurobarometer 76], while only 28% said this in Autumn 2019 [Eurobarometer 92]). These services represented two different conflicting goals: to connect and facilitate information exchange, communication and the social life of people, and also to provide companies with new means of micro-targeting customers with marketing. On these grounds, a battle and critical space of negotiation emerged between the proponents of a big-data business model and a new emerging privacy-by-design business and activist movement (Hasselbalch & Tranberg, 2016). This process generated, among other outcomes, the tougher data protection legal provisions of the European General Data Protection legal reform, which was negotiated between 2012 and 2016.

All in all, the mid- and late-2010s were characterised by a 'battle of systems', the moment of conflict in which technical, legal, cultural and social components of an old and a new system existed simultaneously in a relationship of 'dialectical tension' (Hughes, 1983, p. 79). The 'reverse salients' of BDSTIs and their AISTI evolution were identified, in politics and public opinion in particular, as ethical and social critical problems of the existing systems' data handling and design. That is, 'reverse salients' were approached as *sociotechnical* problems. Accordingly, at the end of the 2010s, not only were engineers proposing and negotiating solutions, but also an increasing number of new types of scientists and experts, combining humanistic studies, social science and philosophy with data science, were participating in negotiations to identify critical problems and propose, among others, applied ethics solutions. These solutions were shaped as responses to the ethical challenges specific to BDSTIs and AISTIs. Therefore, this moment also took form as a negotiation of the societal and ethical values that were to shape the direction of the BDSTIs and AISTIs' technological momentum; that is, their 'mass, velocity, and direction' (Hughes, 1983, p. 15).

Ethics in Policy Vacuums

In 1985, professor of moral philosophy James H. Moor predicted a computer revolution of society. Importantly, the adoption of computers in society will, he argued, 'leave us with policy and conceptual vacuums' (Moor, 1985, p. 272) that will generate particular forms of ethical reflection and value nego-tiation. The societal computer revolution occurs in two stages, Moor proposed, marked by the questions we will ask. In the first 'Introduction Stage', we will ask functional questions: How well do particular technologies function for their purpose? In the second 'Permeation Stage', when institutions and activ-ities are transformed, we will ask questions regarding the nature and value of things (Moor, 1985, p. 271).

I propose to use Moor's depiction of a computer revolution and its policy vacuums in combination with Hughes' theory of sociotechnical change to understand the role of data ethics in governance and policymaking. The policy vacuums that Moor describes present core critical problems and challenges, almost like Hughes' 'reverse salients'. However, rather than problems specific to technical systems, they present the challenges that the introduction of a tech-nology or technological system pose to specific social environments and their established policies, norms and standards. In this way, conceptual muddles and uncertainties are revealed and we are presented with new choices of action (Moor, 1985, p. 266). It is exactly due to this juxtaposition between what we once knew and what we now do not know that we are forced to reflectively consider what we find valuable. That is, the very clash between the techno-logical system (the computer) and existing policy frameworks that we have previously taken for granted will force us to 'discover and make explicit what our value preferences are' (Moor, 1985, p. 267). Said in other words, *ethical reflection* will emerge in situations and moments in which technology alters situations and clashes with existing policies.

We need to acknowledge the ethical compromises we make in these moments as valid components of the governance efforts invested in socio-technical change. They constitute the very cultural compromises shaping the technological momentum. Take, for example, the data ethics public policy initiatives launched in Europe in the second half of the 2010s. They were not solutions to the ethical problems that they explicitly addressed; rather, they represented 'spaces of negotiation' in which values were made explicit and their conflicts were negotiated. These were only the policy initiatives. Several other critical 'spaces of negotiations' emerged in the late 2010s that were particularly critical of the powers of BDSTIs and AISTIs and their main stake-holders, such as the Google employee walkouts to protest against the treatment of women in the company in 2018, or the UK student demonstrations against an automated A-level grading system in 2020. These are the 'ethical moments' that are vital to change: moments in which the norms and values we take for granted are challenged and renegotiated, and alternatives are proposed.

2.   GOVERNANCE

With a 'sociotechnical' analysis, I seek to delineate the combination of not only technical but also social, cultural and economic components that consti-tute the shape of sociotechnical change and, in this context, the role that ethics can play. An apprehension of the complexity of a range of factors is something that we need in order to guide change. As Bijker and Law (1992) put it, tech-nologies do not represent their own inner logic. They are the materialisation of a range of heterogeneous factors, but they are also shaped, even 'pressed' into a certain form that 'might have been otherwise' (Bijker & Law, 1992, p. 3). This is, as I have argued before, an essential view on technological develop-ment and change, such as the evolution of BDSTIs and AISTIs, as it empowers human governance efforts.

We need sociotechnical governance that goes beyond single-sided analyses of the components of the development of BDSTIs and AISTIs in order to effect change. If we want to steer the interests invested in a BDSTI's technological momentum with a human interest, we will not succeed by, for instance, invest-ing only in the development of 'ethics by design' technological components. Neither will we effect change with citizen awareness raising and education alone, or with regulatory requirements and the creation of new systems requirement standards only. We need a distributed governance approach where each of these activities is considered a component of a totality that addresses the opportunities, risks and ethical implications of complex social, cultural and political environments all together.

Furthermore, the 'we' that is doing the shaping is not a single actor that can be identified in just one place (Mueller, 2010; Brousseau & Marzouki, 2012; Epstein et al., 2016; Harvey et al., 2017; Hoffman et al., 2017). Legislators are the most obvious actors of governance, but technological systems are also actors. They also have active socially ordering and governing powers (Reidenberg, 1997; Lessig, 2006), as do users, engineers and designers of technological systems (Winner, 1986; DeNardis, 2012).

We will see in the following discussion of internet governance that the 'governing' of sociotechnical change is a complex heterogenous process. This is also how we should think of the role of 'data ethics' in governance – or what I call 'data ethical governance'. As complex cultural processes of multiple actors and orders, reflexive rather than top-down approaches, that take form in critical moments when the established norms and values of data engineering and design practice and use are challenged by 'untraditional' data ethical problems.

Internet Governance

There can be no infrastructure without some type of governance. Shared frameworks of rule-making, ordering and collective action will always be core to a functioning sociotechnical infrastructure (Star & Bowker, 2006; Bowker et al., 2010). On a very basic technical level, without shared frameworks, the technical components of the system do not interact and the system breaks down or its development is halted. The same counts for legal frameworks, such as laws on how to protect and share data and to protect privacy rights of individuals. They need shared frameworks to function on a very basic level of application. That is, while the negotiation between different values, conflicts of interests and battles of systems in critical moments of a system's development may represent the uncertainty of shared governance agendas, or 'policy vacuums' (Moor, 1985), a well-functioning infrastructure requires a level of agreement to work. This is also what Star and Bowker (2006) refer to as 'handshakes' between the different components. A well-functioning infrastructure is therefore not one in which different frameworks – legal, cultural or technical – are in conflict. It will always represent a compromise or the domination of one standard over others.

The internet is an example of a large-scale information infrastructure that obviously requires institutionally shared global governance through technical – as well as policy and legal – standards to operate efficiently. All the same, in the early 1990s the World Wide Web came into being with a cyber-libertarian imagining of an independent public sphere in which citizens were set free from oppressive state governance by the very decentralised and 'ungovernable' information architecture of the digital network (Mueller, 2010, p. 2). From this, a conception of a different type of bottom-up, people-centred and ethics-based form of governance emerged. As John Perry Barlow famously wrote in his declaration of the independence of the internet in 1996:

> We believe that from ethics, enlightened self-interest, and the commonwealth, our governance will emerge. Our identities may be distributed across many of your jurisdictions. The only law that all our constituent cultures would generally recognize is the Golden Rule. We hope we will be able to build our particular solutions on that basis. But we cannot accept the solutions you are attempting to impose. (Barlow, 1996)

Imaginings like this of technological liberation and freedom from institutional governance lived long into the formative years of the global sociotechnical information infrastructure of the internet (Mueller, 2010, p. 1–13). In the 2000s, however, political battles and negotiations between traditional governments and intergovernmental institutions on the global scene had reached new levels with several regulations and policy initiatives introduced specifically dedicated to the governance of a new internet-based sociotechnical sphere (Brosseau & Marzouki, 2012). The internet was not just a free zone for the 'ungoverned' emancipation of the individual; it was still constituted with a form of governance, although governance of this new public sphere was also not a state-only activity, and increasingly neither was it recognised as such in the official policy sphere.

New actors of governance were emerging and positioning themselves in internet governance policy debates – the engineers and businesses, internet users and their communities (Mueller, 2010; Brousseau & Marzouki, 2012; DeNardis, 2012; Epstein et al., 2016; Harvey et al., 2017; Hoffman et al., 2017). Most significantly, large corporations were designing, but also setting the rules and codes of conduct for their online platforms (Aguerre, 2016; Belli & Zingales, 2017; Franklin, 2019; Jørgensen, 2019; Wagner et al., 2019). That being so, in the early 2000s, multistakeholder governance institutions and initiatives were introduced in the public policy sphere. By way of illustration, the UN Internet Governance Forum (IGF) was formed during the initial World Summit on the Information Society (WSIS) processes in 2003–2005. Here in particular, due to the involvement of civil society and technical community stakeholders, solutions to problems went far beyond the mere technical design of the internet (Brosseau & Marzouki, 2012). Increasingly, for example, human rights issues were included on the official agenda.

Many internet governance studies have focused on the dynamics of the governance of the internet, as well as 'how' to govern a disruptive global and interjurisdictional sociotechnical information infrastructure (Mueller, 2010; Brousseau & Marzouki, 2012; DeNardis, 2012; Harvey et al., 2017; Hoffman et al., 2017; Epstein et al., 2016). As a result, most internet governance scholars move from the perception that the very technological architecture of the internet has brought about new governance models and as such disrupted traditional centralised forms of governance. Internet governance scholar Milton Mueller (2010), for example, describes how the internet as a technological architecture imposes on nation state governance in different ways. The very cross-border communicational technical architecture of the internet means that attempts to impose additional jurisdictional architectures require extra effort. In addition, the architecture of massive information generation, collection and retrieval enables large-scale communication which traditional governments have difficulties in responding to, and which also transforms their governmental processes. Moreover, the decentralised and distributed internet architecture distributes control, and new transnational institutions (such as ICANN) form a new type of power centre for key decisions (Mueller, 2010, p. 4). Ultimately, the internet transforms 'the polity' with new types of collaboration, organisation and mobilisation across borders by converging media and creating new types of communication that lower the cost and empower group action (Mueller, 2010, p. 5). Based on these observations, Mueller uses the term 'governance' rather than 'government' to shift the focus from traditional forms of centralised rulemaking and social ordering steered by nation states. The global sociotechnical infrastructure of the internet has indeed disrupted nation state governance; however, this does not imply that it is not still directed and shaped. All this means is that governance is 'less hierarchical and authoritative' (Mueller, 2010, p. 9).

Internet governance scholars have been particularly preoccupied with the initial official first attempts to negotiate, on an institutional level, a shared global governance approach to the internet and what was also increasingly delineated in global policymaking as the Information Society in the WSIS process initiated in the early 2000s and the following IGF hosted by different countries worldwide every year (Bygrave & Bing, 2009; Mueller, 2010; Flyverbom, 2011; Brosseau & Marzouki, 2012; Epstein, 2013). Increasingly, an approach informed by Science and Technology Studies (STS) has also been used to analyse a more complex set of governance actors and models that emerged in the sociotechnical formation of the internet in society (Epstein et al., 2016). Many different governance components are acknowledged to shape the direction of a sociotechnical system (laws, cultural norms and habits, education, manuals for engineering practices, standards, funding schemes and codes of conduct).

Epstein et al. (2016) delineate the key components of an STS-informed approach to internet governance as based on the following foundational ideas. Firstly, there is a 'plurality' of modes of governance that are also taking place and being enforced in a variety of fora and according to a diverse set of 'normative systems' from law and technology to social practices. Then, the technical infrastructure has 'nonhuman' agency that not only orders the social but also controls it. Moreover, it is not only the official and extravagant actions (such as regulations and political agendas) of humans that 'govern'. It is also the invisible 'mundane practices' of humans that shape the 'design, regulation and use of technology'. Importantly, a key focus is 'controversies as structuring and performative processes' where different stakeholders' interests are juxtaposed and negotiated, exposing their different notions of governance. Finally, the notion of 'multistakeholderism' is brought forward by an STS-informed approach when acknowledging the many actors that participate in 'doing internet governance', and specifically in the role of private actors (from users to industries) in the context of decision-making and governance of the internet (Epstein et al., 2016, p. 6–7).

The term 'reflexive coordination' is introduced by Jeanette Hoffmann et al. (2017) in an attempt to embrace these heterogenous components of internet governance:

> Our approach on governance proposes a fundamental shift in perspective: instead of gradually extending a regulatory perspective beyond nation-states, public decision-making and formal policy instruments, we suggest studying Internet governance as a continuous heterogeneous process of ordering without a clear beginning and endpoint. (Hoffman et al., 2017, p. 1412)

This is a type of governance that takes place 'in-between' the top-down, intentional steering of states and the heterogeneous disruptive and less organised ordering activities of dispersed (new and traditional) actors of internet governance (the state, engineers, users, citizens, scientists and technological artefacts), considering their intertwined intentional and unintentional 'multiple orders' (Hoffman et al., 2017, p. 1410). By focusing on the way in which different actors are coordinated and interrelated, the complexity and diversity of different actors of governance are brought to light. Their coordination activities might be simple and ordinary and – on the face of it – unexciting. Nevertheless, they do create a type of social order (Hoffman et al., 2017, p. 1413). This type of routine and habitual coordination of order can only be steered reflectively ('reflexive coordination') in critical moments when different norms, assumptions and understandings of situations clash (Hoffman et al., 2017, p. 1415).

I wish to use these reflections on governance from a field of internet governance studies to explore the role of data ethics in the 'governance' of the sociotechnical change of BDSTIs and AISTIs. They are descriptions of new legitimate modes of governance emerging as a response to, but also reinforced by, the specific architecture of the internet. It is a type of steering of sociotechnical change that is institutionally engineered as well as nonengineered in cultural practices of, for instance, engineers and users. Governance can also be understood here as 'open-ended' in the sense that it does not have a predefined start and end point or solution that we can steer towards. Even a law reform is not only a clearly delineated negotiation process that takes off with a proposal and ends with the adoption of a new law; it also consists of follow-up evaluation mechanisms and other forms of intervention (Brøgger, 2018).

Here, I also want to combine the critical moments in which simple coordination activities transform into 'reflexive governance' (Hoffmann et al., 2017) with Hughes' moments of conflict in a technological system's development, where 'reverse salients' are identified as critical problems of the system and dialectical tension between different systems occurs (1983, 1987). They are

also ethical situations, as Moor describes, where problem-solving and negotiating actors are more focused on conceptual activities and explication of
the nature and value of things due to 'conceptual muddles' and uncertainties
of policy frameworks caused by socially disruptive technologies, such as
the computer (Moor, 1985, p. 266). That is, they are the ethically reflective
moments that emerge out of relations, as Rorty describes, when 'controversy
arises' (Rorty, 1999, p. 73), and thus, they are the essence of the type of 'data
ethical governance' I want to propose as the governance component of a data
ethics of power. This is when standard world views, norms, and data practices
and cultures clash and force out a particular type of reflection of the social construction of our assumptions; and accordingly, an ethical reflection on alternative data policies and practices emerges (Moor, 1985). Or in other words,
it is when the established norms and values of data engineering and design
practices and use are challenged and 'untraditional' data ethical problems
are revealed that new policies, strategies and solutions are formulated. These
critical cultural moments (which I will describe in more detail in Chapter 4)
take form as values-based governance and are characterised by the inclusion of
multiple actors with an interest in the data of BDSTIs and AISTIs. This is also
the way in which Rainey and Goujon (2011) describe 'ethical governance', as
a reflexive rather than top-down approach that takes into account the conditions for ethical reflection in particular:

> What's required is an approach that can offer first criteria of evaluation and second
> a more interesting way to address the conditions not only for an ethical reflexivity,
> but also for determining the conditions of construction of ethical issues, of ethical
> norms, and the conditions for their adoption and implementation. (Rainey &
> Goujon, 2011, p. 54)

Sustainability and The Data Pollution Problem

At Google's annual developers' conference in 2017, CEO Sundar Pichai reit-
erated the company's 'AI-first' mission to make machine learning an umbrella
integrated into all Google platforms to enhance all services from video, search
and email to mobile.[1] In a *Fast Company* interview, Pichai remembers this
approach to put AI first in all Google products as a moment of existential revelation: 'This thing was going to scale up and maybe reveal the way the universe
works. ... This will be the most important thing we work on as humanity'
(Brooker, 17 September 2019).

An all-embracing and dedicated AI approach like this was, in the early 21st
century, not unique to Google. It was evident in BDSTI and AISTI practices
in general that were developed and adopted with a sense of urgency similar
to the urgency of the big data imagination of the 1990s. BDSTIs were in this

way also transforming into AISTIs with advanced technical data systems designed to sense a datafied environment in real time, training with big data and learning from it to evolve autonomously or semi-autonomously. AISTIs had social components, facilitating and increasingly constituting the public and private spheres, and they were already in part institutionalised in systems requirements standards for IT practices and regulatory frameworks for data protection. However, as illustrated throughout this book, towards the end of the 2010s, concerns regarding the sustainable values of AI and big data, the ethical implications of autonomous AI systems, and the adverse social impacts of big data, were increasingly also being raised in society and creeping into engineering communities. Thus, in 2019, at Google's annual developers' conference, Jen Gennai, Google's head of responsible innovation, told a crowd of developers: 'We've identified four areas that are our red lines, technologies that we will not pursue. We will not build or deploy weapons. We will also not deploy technologies that we feel violate international human rights' (quoted by Brooker, 17 September 2019). Not all of the audience was convinced about Google's ethical approach to AI though. As one participant said to the journalist present at the event 'I don't feel like we got enough. … They are telling us, "Don't worry about it. We got this." We all know they don't "got this"' (quoted by Brooker, 17 September 2019).

If we want to govern the direction of the sociotechnical evolution of BDSTIs and AISTIs, we have to take into account a complex intertwined network of relations. This includes the different worldviews and imaginations driving forward particular scientific and technical developments. The doubts of a Google developer conference participant as to Google's good intentions are symptomatic of a moment of controversy, the 'battle of systems' of the late 2010s, which constituted a crucial uprooting of what had previously been taken for granted in AI and big data technology practice, including the dominant invested values. Thus, while big data and AI-first 'mindsets' and ideas were still driving the development of BDSTIs and AISTIs, these were increasingly also challenged by other 'values'-based approaches, such as 'privacy by design', 'ethical design', 'human-centric', 'trustworthy AI' and 'sustainable AI'.

I want to argue that we can take a moment like this, the most critical moment, where foundational values are explicated and negotiated, and look at it as a valid component of governance. This is the moment when new policies and directions for the evolution of sociotechnical systems are created. The values that we formulate explicitly in response to a controversy or crisis like this, often with great social force, concern who we are and where we are going. This is also when ethics play a crucial role as an explicit cultural reflection of human values that are negotiated with other competing interests. The most important sociotechnical changes take place when these cultural values are translated into concrete technical solutions, science, forms of innovation, cultural movements and policies. However, this is in no way a straightforward process, and it is also a slow process, taking form sometimes over decades with the involvement of multiple agents and actors.

Think, for example, about the way in which the concept of 'environmental sustainability', articulated in response to the pollution of our physical environment, took form over the last 50 years. Think of how these values of sustainability and concerns regarding our planet's future and intergenerational justice became drivers for entire new legal and policy frameworks, such as the national and international environmental laws and collaborations introduced in the 1970s (like the first UN Conference on the Human Environment in 1972). Environmental concerns and values of sustainability transformed industries, for instance the car industry, and drove forward the development of new industries and sciences, such as 'green tech'. Today, companies need an 'environmentally friendly' profile and conduct, not only because they must adhere to environmental laws, but because taking care of the environment and acting in a sustainable manner as a company is just sound business practice. It is an investor demand, a legal requirement, and a demand among customers and in society in general.

Now, consider the impact of digital technologies on the environment. Firstly, the impact on our natural environment of digital technologies, data storage and processing are tremendous. In 2019 their share of global greenhouse gas emissions was estimated to amount to 3.7% (The Shift Project, 2019). Data centres, for example, accounted for 1% (and steadily growing) of total global electricity demand. The majority of this growth was attributed to cloud computing by the largest big data companies such as Amazon, Google and Microsoft (Mytton, 2020). Looking at data-intensive technologies, such as AI, the impact is evidently also significant. For example, when training (including tuning and experimentation) a large AI model for natural language processing, such as machine translation, the carbon cost is seven times the carbon footprint of an average human in one year (Strubell et al., 2019; Winfield, 28 June 2019).

Importantly, I also want to include here the impact of big data on our social and personal environments. The computer security and privacy technologist Bruce Schneier has used the term 'data pollution' to describe the privacy implications of big data technology and systems. He sees this as a core environmental problem of our age:

> this tidal wave of data is the pollution problem of the information age. All information processes produce it. If we ignore the problem, it will stay around forever. And the only way to successfully deal with it is to pass laws regulating its generation, use and eventual disposal. (Schneier, 6 March 2006)

We therefore can, and should, think of 'data pollution' as the interrelated adverse effects of big data on natural, social and personal environments. Yet, 'environmental awareness' in society and among the companies responsible for the 'data pollution' of this age were, in the early history of BDSTIs and AISTIs, lagging greatly behind other forms of environmental concerns.

A company's environmental awareness and sustainable values generally concern the impact of its practices on otherwise healthy ecosystems whose survival depends on sensitive balances between many different components. Since the 1960s, awareness of the impact of the pollution of natural environments by various industries – such as transportation, manufacturing and energy – has gradually been transformed into legal requirements, technical standards and forceful social demands. Companies are, for example, today required to make use of energy labels, live up to ECO design standards, and to monitor and systematically improve their environmental performance, while customers are demanding 'eco-friendly' products. However, these same 'environmental' mitigation tools – social and legal requirements – are yet to be developed to mitigate the adverse environmental impact of 'data pollution'. Generally we need a better understanding of the constitution and balance of our 'data eco-systems'.

Let us think about a couple of examples here. In 2014 data scientists at Facebook conducted a large-scale experiment on 689,003 randomly selected Facebook users by filling their newsfeeds with positive or negative stories and then measuring their emotional reactions (Kramer et al., 2014). When the story surfaced in the public sphere there was a public outcry, and a Facebook spokesperson immediately stepped forward with a public apology. However, the apology did not concern the 'environmental impacts' of the data experiments; that is, it did not concern an ethically questionable company use of data, such as the manipulation of personal lives of unknowing users, or the data protection legal implications. In fact, she was only apologizing for the fact that the experiment had been 'poorly communicated'. As she said, these experiments with users' data were just business as usual: 'This was part of ongoing research companies do to test different products, and that was what it was' (Krishna, 2 July 2014). After several years of revelations similar to this regarding the ethically and legally questionable use of Facebook users' data, it is today indisputable that company data science practices such as these are ethically problematic, are certainly also debatable in legal terms, and most definitely their adverse effects on the personal environments of users can easily be identified due to years of investigative efforts by critical data journalists and scientists. Thus, social demands and responses to revelations like these have also grown in force. Nevertheless, we still do not have the same governance tools to mitigate these environmental risks as we do when addressing other more traditional environmental concerns.

We can here take another example of a public revelation of a company's pollution of the environment, but this time from an industry that had to respond to more established environmental legal frameworks. In 2015, the car company Volkswagen was caught in deploying sophisticated software to cheat emissions tests and allow its cars to produce up to 40 times more pollution than allowed. This was in public discourse considered a big scandal with global ramifications. Not only had the company caused more pollution, impacting our natural environment, it had also manipulated data to be able to do so. The car company was immediately forced to recall hundreds of thousands of cars; in one day €15 billion was wiped off the company's share price on the stock exchange. A Chief Executive went out publicly with a big apology and later resigned, and governments around the world called for action (Topham, 25 September 2015).

Let me here rewrite parts of the article in *The Guardian* that described the Volkswagen scandal in order to create an illustrative fictional example of what a similar response to a 'data pollution' scandal of the imaginary big data company DD Mobile could look like in the future:

> DD Mobile has been told to recall 482,000 devices in Europe after it was caught deploying sophisticated software to illegally surveil users and allow their devices to harvest up to 40 times more data than allowed. The newly established European Data Protection Agency claims DD Mobile installed surveillance software in their devices. The EDPA says: 'We intend to hold DD Mobile responsible. We expected better from this company. Using a surveillance software in devices that evade data pollution prevention and data protection standards is illegal and a threat to privacy.' The EDPA warns that DD Mobile will be further investigated and could face other action for breaching the Data Pollution Directive and the General Data Protection Regulation, including a maximum fine of up to €37,500 per device, or €18 billion. (My rewrite of Topham, 25 September 2015)

This example addresses fictional institutions and legal repercussions and I use it in the context of what we know and have been accustomed to when it comes to other environmental scandals to illustrate the concept of 'data pollution'. It is evident that with the emerging social awareness of the data pollution of AISTIs and BDSTIs we will also see a transformation in the way in which policymakers and consumers address the data pollution and the sustainable values of technology companies. The legal scholar Omri Ben-Shahar, for example, describes the development of 'an environmental law for data pro-tection' to mitigate the effects of data pollution with legal tools similar to the ones created to control other forms of industrial pollution (Ben-Shahar, 2019). However, the response will not be legal only. It will be cultural and social. We will increasingly see and feel the impact of the 'data pollution' of big data technologies on our social and natural ecosystem and on future generations, with adverse effects on everything from our privacy and our democracies to the carbon footprints of the data exhaust of big data technologies. We will respond by translating these concerns into law, design, science and education. Thus, the 'data pollution' of the very data design, storage and processing of a data technology, product, service and company, will also evolve into a key area of environmental concern for the companies and institutions responsible for this particular kind of pollution. Change will indeed happen, but not in one day, because before we can mitigate an adverse environmental impact, we need to also see it. As the robot ethicist Aimee van Wynsberghe says, this is why we need a movement today that considers sustainability not only as a goal of a technology, such as AI, but addresses in concrete terms the very sustainability *of* developing and using these technologies (van Wynsberghe, 2021).[2]

Data Ethical Governance

The technological momentum required for a large sociotechnical system to consolidate in society is not just an arbitrary composition of social, economic and cultural factors mixed together by an inexplicable will of nature; it has a shape that guides the direction, values, knowledge, resources and skills that form the technological architecture of the system – its governance, adoption and reception in society. At times, as I have tried to illustrate with the example of the evolution of 'environmental and sustainability concerns' and 'data pollution', this shape is more explicitly cultural and values-oriented than others. This 'cultural' shape, the reflective ethical evaluation and value-orientation, is a moment we have to think of as not just a momentary critical response to critical problems. It is also a valid component of governance. In the 2010s, an emerging awareness of the adverse effects of data pollution were translated into new forms of innovation, laws, science and intergovernmental collaborations. 'Data ethical governance' was thus also increasingly recognized as a component of governance in public policymaking, where 'data and AI ethics policy initiatives' were accepted as components of institutionalised forms of governance.

Winfield and Jirotka (2018) use the term 'ethical governance' to present a case for 'a more inclusive, transparent and agile form of governance for robotics and artificial intelligence (AI) in order to build and maintain public trust and to ensure that such systems are developed for the public benefit' (Winfield & Jirotka, 2018, p. 1). 'Ethical governance', they argue, goes beyond just good and effective governance, it is 'a set of processes, procedures, cultures and values designed to ensure the highest standards of behaviour' (Winfield & Jirotka, 2018, p. 2). Governing the development of robotics and AI with an ethical framework, they therefore argue, requires a diverse set of approaches, from those at the level of individual systems and application domains to those at an institutional level (Winfield & Jirotka, 2018, p. 2).

In continuation of the previous discussion of 'internet governance' as a multi-actor and agile process and here of 'ethical governance' as a set of activities and approaches designed to ensure the 'highest standards of behaviour', I also want to emphasise an applied ethics of governance approach informed by pragmatism. That means that I consider 'ethical governance' not only a negotiation of and application of foundational ethical values, but also in terms of the very conditions, practices and processes that produce ethical reflection.

If we take, for example, the role of ethics in the governance of AISTIs in the late 2010s, we will see that there were two different forms of 'applied ethics' in motion and accordingly also two different approaches. The most visible place we could look for ethics in the public governance debate in this period would be in the overwhelming amount of normative AI ethics guidelines and principles produced by various state, intergovernmental, civil society and industry stakeholder interest groups worldwide (Fjeld et al., 2019; Jobin et al., 2019). This was also where the main critique of the role of ethics in the governance of AISTIs and BDSTIs at first was applied. It was not clear how these high-level meta-principles could be implemented and translated into practice, nor whose interests they represented. Would they make any difference to the standards of practice for the global community of practitioners, developers, users and policymakers? To move the dialogue on principles forward in a more constructive direction, one response was therefore to trace thematic convergences between the various documents and in this way create a common normative framework with a set of universal principles (Floridi et al., 2018; Winfield & Jirotka 2018). However, here I want to illustrate how another pragmatist applied ethics approach could be, and actually also was, applied, at the same time. It was less visible in the public debate about ethics, as it was not presented as 'ethics' per se; yet it was just as significant, if not more so, for the ethical governance of AISTIs and BDSTIs.

Professor of philosophy and politics Andrew Altman describes a pragmatist applied ethics that involves a 'contextualist view of justification' (Altman, 1983, p. 232). This means that any ethical assumption (which also includes high-level ethical principles) can be challenged in context – or, in other words, an ethical theory or approach can only be justified in practice. Rather than the development of ethical normative frameworks, we should therefore think of applied ethics as a practice and reflective process. To begin with, this means that we would have to consider all the different ethics guidelines produced in terms of their unique points of reference, their contexts of application and their 'non-neutral' ethical points of departure. However, it also means that 'ethics' is to be found in not only the very negotiation of foundational values, the creation of normative ethical theories or frameworks, but also, and more importantly, in the very activities that are testing these types of ethical ideas in practice. As follows, let us now explore in more detail 'data ethical governance'. Not as a set of ethics principles and guidelines, but as the actual activities of a particular moment that are set in action in order to consciously test an ethical idea. These are the activities that I consider the applied ethics of a data ethics of power.

In the 2010s, a variety of stakeholders from the civil society, policy, business and technology fields set in motion various activities with what one could consider a shared ethical idea: to develop 'human-centric' data cultures in response to the ethical implications of the dominant data cultures of data design and practice. I explore some of them here.

• Data Ethics Initiatives

I have previously described the emergence of public policy initiatives in Europe with explicit reference to data ethics. These appeared alongside several civil society, academic and technology initiatives in which data's ethical implications were framed as issues of a growing data asymmetry between big data institutions and citizens in the very design of data technologies, and solutions sought along these lines. As an illustration, the conceptual framework of the 'Personal Data Store Movement' (Hasselbalch & Tranberg, 27 September 2016) was described by the non-profit association MyData Global Movement as one in which individuals 'are empowered actors, not passive targets, in the management of their personal lives both online and offline – they have the right and practical means to manage their data and privacy' (Poikola et al., 2018).[3] Here, the emphasis was on moving beyond mere legal data protection compliance to implement values and ethical principles such as transparency, accountability and privacy by design (Hasselbalch & Tranberg, 2016). In particular, mitigation of the ethical implications was sought with values-based approaches to the design of technology, as called for by one of the key people behind the ethical standards movement John C. Havens in his book *Heartificial Intelligence* (2016): for example, engineering standards – such as the P7000s series of ethics established by IEEE, one of the world's largest engineers' organisations – and AI standards that strove towards developing ethics-by-design technical requirements standards for the development of AI.[4]

• The Privacy Civil Society Movement

A crucial component of the data ethics momentum of the 2010s was the many consumer and citizen awareness initiatives launched in different civil society contexts with reference to the online power asymmetries between citizens, states and private industry. In the 1990s, the privacy movement was dedicated to the development of privacy-enhancing technologies (PETs) in technically savvy communities, with the introduction of the TOR anonymity software and the TOR project and movement, among others (Hasselbalch & Tranberg, 2016, p. 85). However, in the 2010s the privacy movement was starting to take a more popular form, with organisations such as the UK-based Privacy International and US-based Electronic Frontier Foundation (EFF) that dedicated campaigns to citizen awareness of online privacy with regard to state surveillance practices in particular. The crypto party movement, initiated by the Australian journalist Asher Wolf in 2012, resulted in a range of self-organised crypto parties worldwide, which citizens could attend to learn how to protect their privacy and anonymity online. Increasingly the popular privacy movement was also taking into account the big data practices of the private industry, offering 'digital self-defence' tools and alternatives to the big data technology industry giants' consumer services (Tranberg & Heuer, 2013; Hasselbalch & Tranberg, 2016; Veliz, 2020).

- **Ethics by Design and Critical Investigations of Data Systems**

A specific applied ethics focus on technology and design was spelled out in numerous 'ethics by design' activities. 'Ethics by design' is a term used to address the design and design practices of a technology (Dignum et al., 2018) (I return to the 'ethics by design' and VSD approach in the next chapter). Here, it is important to mention 'privacy by design', which was originally developed by the former Canadian Information and Privacy Commissioner Anne Cavoukian (2009), as it specifically focuses on organisational and design practices that seek to embed 'privacy' as a value in the data design of a technology. In addition, we may also include what the philosophy and technology scholar Philip Brey describes as a 'Disclosive Computer Ethics', which seeks to identify and reveal ethical implications in opaque information technologies (Brey, 2000, p. 12). As follows, case studies of specific data processing software have been crucial to 'data ethical governance'. Examples are the 'Machine Bias' study (Angwin et al., 2016), which exposed discrimination embedded in data processing software used in US defence systems; Gender and African American Studies Scholar Safiya Umoja Noble's (2018) investigation of Google's discriminatory search algorithms; or the mathematician Cathy O'Neil's (2016) analysis of the social implications of the math behind big data decision-making in everything from obtaining insurance and credit to getting and holding a job. Here, the human rights lawyer and Director of the Ada Lovelace Institute, Carly Kind, describes three waves of 'ethical AI' where the two first waves were focused on high-level principles and technology as the solution to ethical problems, and the third wave is finally as she says 'starting to mean something' by exploring questions of power, equity and justice (Kind, 2020)

- **The Law**

A range of legal studies has critically assessed the legal framework for big data in terms of sociotechnical development, privacy (Solove, 2006; Cohen, 2012), human governance, and the implementation of the rule of law and human rights in autonomous data-based systems, as well as AI and robotics (Pasquale, 2015, 2020; Hildebrant, 2016; Latonero, 2018; Nemitz, 2018; Smuha, 2020). Many legal studies have focused on the legal framework of the European general data protection regulation, GDPR (for example, Wachter et al., 2017; Zarsky, 2017; Wachter, 2019). In the context of legal instruments to regulate power distribution, it is here relevant to note a distinction in scope and logic between privacy and data protection, as Paul De Hert and Serge Gutwirth (2006) notice: that is, privacy used as a 'tool of opacity' to stop or set 'normative limits' to power, and data protection as a 'tool of transparency' to channel 'legitimate power' (De Hert & Gutwirth, 2006).

There have also been various emphases on specific challenges in law with regard to children's personal data (Hof et al., 2019), for example, or in the context of smart toys (Keymolen & Hof, 2019). Here, it is also relevant to mention work on a legal framework for 'data trusts' presented by Sylvie Delacroix and Neil D. Lawrence (2019). They consider the development of a plurality of 'data trusts' that individuals can choose between – an empowering alternative to what they refer to as a '"one size fits all" approach to data governance', since that will allow 'data subjects to choose a Trust that reflects their aspirations, and to switch Trusts when needed' (Delacroix & Lawrence, 2019, p. 236).

- **The Discourse**

Critical data studies have deconstructed the cultural narratives of dominant data cultures of institutions, industries and communities that design and build the systems in which big data is processed and analysed (from social networking services to AI agents; Bowker & Star, 2000; Kitchin & Lauriault, 2014; Albury et al., 2017; Acker & Clement, 2019). Surveillance studies have investigated the words that either empower or disempower us when we discuss privacy, rights and democracy in the context of, for example, security and the technology we adopt to mitigate perceived risks (Lyon, 2014), and legal studies have investigated the discourse of law (Solove, 2001, 2002, 2008; Cohen, 2013). As professor of communications, Klaus Bruhn Jensen illustrates the way we communicate and collectively reason about the 'common good', 'ethics' and 'justice' of a society is translated into human practice, action and social relations in very concrete ways (Bruhn Jensen, 2021). Thus, the counter-narratives that feed into the development of alternative data cultures are crucial. In 2014, when Tranberg and I began to discuss and research for our book on data ethics as 'a new competitive advantage', this was in fact one of the things we wanted to do: to trace and present an alternative narrative to a then-dominant discourse that values such as privacy were outdated and an obstacle to innovation (Hasselbalch, 2013, 2014). Back then, many were not convinced that data ethics was a term that would appeal to companies or suspected that they would tell us that we simply did not understand innovation. Nevertheless, by the end of the 2010s the 'competitive advantage' discourse was in fact overturning other 'big data discourses' in public debate and in policy. The first data ethics group established by the Danish government was, for example, created with the specific objective to turn data ethics into a competitive advantage for Denmark. In addition, with a profound impact on public discourse in the late-2010s, Shoshana Zuboff deconstructed the dominant meta-narrative of the Big Data Society (2014, 2016, 2019) as one that was defined by powerful industries and presented what she referred to as an alternative 'synthetic-declaration' that 'value[s] people, and reflect[s] democratic principles' (Zuboff, 2014).

Spaces of Negotiation

In our daily lives, infrastructures are mundane things. We take them for granted. The streets we walk on, the bridges we cross most often have no visible being as spaces of moral and ethical compromise. Yet, when they break down or malfunction, their embodied politics becomes visible. This moment, where the narrative (as Susan Leigh Star, 1999, calls it) – or what could also be referred to as the politics (as per Langdon Winner, 1980) – of an infrastructure becomes visible, is also the key moment that will, based on the negotiation of interests that follows, give shape to the direction of the infrastructure's transformation, or in other words, to sociotechnical change.

Data ethical governance takes place in these moments of controversy in what I have called 'spaces of negotiation'. I argue that data ethical governance has a function in creating spaces of sense-making and negotiation that happen in time, in a moment of crisis, just before their consolidation.

In the early 21st century, BDSTIs and AISTIs were rapidly consolidating in our public and private spheres. At the same time their social and ethical challenges became increasingly visible. Most of us remember Cambridge Analytica and the Snowden revelations, but also increasingly we either ourselves had personal experiences, or heard about other people's personal stories, where individual lives were clashing with the predictions and the decisions of an arbitrary data system. The creepy knowledge that a service suddenly had about you, the man who was arrested based on an erroneous match made by a facial recognition system, the student who got a bad grade from an algorithm. As a result of these encounters with the ethical and social implications of sociotechnical data systems, their 'politics' and 'values' were also increasingly often questioned in policymaking and public debate and alternative technical designs and business models, as well as legal and social requirements, were introduced.

The first function of data ethics in governance can be found in the 'spaces of negotiation'. They take form in what, in Chapter 4, I call 'critical cultural moments' when controversy arises and different human values, cultures and reflections are pulled to the foreground and are renegotiated due to a disturbance of 'the state of affairs'. The 'data ethics spaces of negotiation' are formally introduced and framed in policy processes (as was the case with the 'data ethics public policy initiatives' that I have previously described). However, they also happen informally in micro-settings of policy work and they include very concrete discussions about specific values. As, for example, one policy advisor to a member of the European Parliament said to me, when describing the role of ethics in the GDPR negotiations:

> The moment you see a conflict of interest, that is when you start looking at the values ... normally it would be a discussion about different values ... an assessment of how much one value should go before another value ... so some people might say that freedom of information might be a bigger value or the right to privacy might be a bigger value. (Interview, Internet Governance Forum 2017, Hasselbalch 2019)

Data ethics spaces of negotiation also more and more often comprise existential reflections about the general evolution of society. As a country representative in the Committee of Ministers of the Council of Europe once said to me: 'We need to slow down a little bit and to think about where we are going' (Interview, Internet Governance Forum 2017, Hasselbalch 2019).

NOTES

1. See Sundar Pinchai's talk at Google's 2017 annual conference: https://events
   .google.com/io2017
2. See also the Data Pollution & Power White Paper (2022) and the Data Pollution &
   Power Group mini reports 2021–2022 www.datapollution.eu
3. See also the white paper 'MyData – An Introduction to Human-Centric Use of
   Personal Data' (2020) www.mydata.org
4. See Ethics in Action, P7000s standards. https://ethicsinaction.ieee.org/p7000/

3.   Artificial Intelligence Socio-Technical Infrastructures (AISTIs)

'The sad thing about artificial intelligence is that it lacks artifice and therefore intelligence.'
Jean Baudrillard, 1983

Artificial Intelligence (AI) is everywhere, and it is nowhere, because what do we actually mean when we talk about AI? Is it a sophisticated improvement of our outdated human software? Is it a sci-fi scenario where an out-of-human-control machine out-competes humankind? Or, is it a commercial trade secret? Words are very powerful and, as abstract as they might seem sometimes, they actually have real consequences. Real laws are implemented based on the particularities of language; real business decisions are made; and real people's lives are affected by the specific use of words and the worlds they portray. Evidently, the way we talk about AI defines what we think we can do with it and ask from it.

The founder of the singularity movement, Ray Kurzweil, believes that AI is the next step in human evolution:

> Biology is a software process. Our bodies are made up of trillions of cells, each governed by this process. You and I are walking around with outdated software running in our bodies, which evolved in a very different era. (Lunau, 14 October 2013)

The late scientist Stephen Hawking considered the power of AI an uncontrollable autonomous force:

> The development of full artificial intelligence could spell the end of the human race. … It would take off on its own, and re-design itself at an ever-increasing rate. Humans, who are limited by slow biological evolution, couldn't compete, and would be superseded. (Cellan-Jones, 2 December 2014)

The co-founder of Google, Larry Page, on the other hand, sees AI as just another (Google) service:

> Artificial intelligence would be the ultimate version of Google. The ultimate search engine that would understand everything on the web. It would understand exactly what you wanted, and it would give you the right thing. We're nowhere near doing

that now. However, we can get incrementally closer to that, and that is basically what we work on. (Marr, 2017)

Whatever we say AI wants to be or can do for us will shape the role it plays in society (Hasselbalch, 2018).

In the mid 2010s, the term AI gained traction in public discourse, and particularly in business and technology companies that started rebranding their big data efforts as AI (Elish & boyd, 2018). Concurrently in global policymaking, AI became a new item on the agenda of nations and intergovernmental institutions with the dedicated development of policy and investment strategies. With no shared definition, the term first and foremost was used generically to describe a sociotechnical evolution of big data technological systems. Amplified computer power and the vast amount of data generated in society had empowered machine learning technologies to evolve and learn to recognise faces in pictures (pattern recognition in images, 'facial recognition'), to recognise speech from audio (pattern recognition in audio, 'voice recognition'), to drive a car autonomously (rendering objects in an environment and performing a risk assessment), and to understand individuals when micro-targeting services and information ('profiling' and 'personalisation'). These were all practical applications of AI systems increasingly adopted by companies and states to not only solve simple problems, and analyse and streamline disparate data sets, but also to act in real time, sensing an immediate environment and supporting critical human decision-making processes.

In this chapter, we will examine the history and special characteristics of BDSTIs with AI capabilities – what I also refer to as AISTIs, their ethical implications, and the ethical theories that address these implications. The core objective of this chapter is to narrow down the data ethics of power considerations specific to AISTIs.

1.   CAN A MACHINE THINK?

Humans creating intelligent machines or life out of inanimate or dead things has been a narrative throughout human history, from the Greek myth of Deucalion and his wife Pyrrha, who made beautiful people by throwing stones over their shoulders, to literary and filmic depictions of the living corpse Frankenstein, the string doll Pinocchio, Metropolis' humanlike machine, Maria – and the first autonomous car, Herbie, in the 1968 film in which the character Tennessee Steinmetz says to his friend, the owner of the car:

> *Jim, it's happening right under our noses and we can't see it. We take machines and we stuff 'em with information until they're smarter than we are. Take a car. Most guys spread more love and time and money on their car in a week than they do on*

their wife and kids in a year. Pretty soon, you know what? The machine starts to think it is somebody.

Yet, the very scientific conception of a computation process with intelligence was most famously theorised by the mathematician and computer scientist Alan Turing, who in 1950 developed a method for testing a machine's ability to display intelligent behaviour indistinguishable from that of a human (Turing, 2004).

The term Artificial Intelligence was, however, first coined in 1956 by the mathematics professor John McCarthy at the Dartmouth Summer Research Project seminar. He wanted to shift the focus of attention of computer scientists and mathematicians in the field of computation processes from the mere automation of these to the 'intelligence' of computers (Moor, 2006). Could a computation process do more than just process information and actually *think* information and learn from it like a human?

In the early AI research field, AI was explored by discerning the key differences (and similarities) between the human brain, feedback systems and digital computers (Crevier, 1993). As evidence of the similarities with the human mind and potential superiority of AI, chess-playing computer systems, for instance, were later developed, capable of creating and acting according to game strategies. The most famous example was IBM's Deep Blue, which became the first computer to beat a chess champion when it defeated Russian Grandmaster Garry Kasparov.

But fifty years after the Dartmouth seminar, when five of the original scientists of the first seminar reconvened with other key people in the evolving and increasingly interdisciplinary field of AI research to discuss the next fifty years of AI development, the ambitions of the early AI researchers were more disparate (Moor, 2006). While McCarthy was now less convinced about the creation of human-level AI, others imagined AI with feelings and affectations, and the scientist and founder of the Singularity movement, Ray Kurzweil, was certain that a Turing test-capable AI was not far away. The social science and psychology scholar Sherry Turkle, on the other hand, was less interested in the future potential of the intelligence of machines and more concerned with the human implications (Moor, 2006). One could propose here that Turkle represented a general twofold humanistic concern with the endeavour of the strand of AI science that sought to replicate the processes of the human brain and create intelligent nonhuman agents. Firstly, the relations between humans and machines alters human societies and minds in profound ways (Turkle, 1997). Secondly, we may add that the foundational questions regarding a computers' intelligence and the undertaking to develop computer intelligence, and even consciousness, have from the outset been intertwined with concerns regarding what it means to be human and our unique status as the centre of our environments. Is the human neural system just another information processing system, complex, but also as material as the data processing of a machine? (Wiener, 1948/2013; Bynum, 2010). And if this is the case, is it even possible to argue that the human data processing agent ('inforgs', Floridi, 1999) has rights that other nonhuman agents (also 'inforgs', ibid. 1999) in our information environment ('infosphere', ibid. 1999) do not have? As Steve Woolgar puts it:

> Attempts to determine the characteristics of machines are simultaneously claims about the characteristics of non-machines. ... In discussing and debating new technology, protagonists are reconstructing and redefining the concepts of man and machine and the similarity and difference between them. (Woolgar, 1987, p. 324)

The term AI has gone through several societal and scientific stages representing different aspirations to create human-level AI or just computers with very advanced problem-solving capabilities. In 1980, the philosopher John Searle famously illustrated this fundamental conflict of views on the capabilities of AI in his Chinese Room example. He imagines that he is locked in a room where he is to respond to Chinese characters slipped under the door by following a computer program on how to do this. He does not understand Chinese, but by doing just what a computer does, following the program for handling the Chinese symbols, he can respond and slip back correct Chinese characters under the door, which convinces the ones outside the room that there is a Chinese speaker in the room. This example, he argues, illustrates the inadequacies of the Turing method. A computer may indeed create a satisfactory response if it is programmed to act according to the rules for interaction, but this does not mean that it is capable of understanding. Searle himself would not leave the room with an understanding of what was communicated to him through the door or what he responded himself. He therefore concludes from this example that strong AI has 'little to tell us about thinking, since it is not about machines but about programs, and no program by itself is sufficient for thinking' (Searle, 1980, p. 417).

Searle's argument illustrates a conflict in the original aspirations of AI research to create respectively machines that think and understand by themselves or machines that 'just' process information and solve problems for humans. It also represents the early outlines of different sets of discourses that later would form essential frameworks for the development of AI research and its adoption in society. Elish and boyd (2018) describe AI as a technology that has always been suspended between the real and the imaginary cultural perceptions, one being that of the agent machine that acts outside of human control:

> Western perceptions of what AI is – what it can and cannot do, and what it might yet do – are informed by long-standing cultural imaginaries of machines that escape the

control of their creators, and the promises and perils of automata and artificial life. (Elish & boyd, 2018, p. 8)

In the 2010s, the idea that machines may one day evolve entirely autonomously out of human control was still thriving; for example, Stephen Hawking warned in 2014 that the development to full AI could end mankind (Cellan-Jones, 2 December 2014); moreover, the founder of the Singularity movement, Ray Kurzweil, predicted in 2016 that AI will update the outdated software of humanity to create an entirely superior intelligence (Lunau, 14 October 2013). These ideas can also be traced in more general public discourses on AI intelligence and potential agency of machines, and accordingly in imaginings about the imminent potentials or threats of AI. Concerns regarding AI agents that replace the human labour force or the artistry and creativity of new AI systems represent the imagining of an autonomous new agent in society.

As such, it may be argued, as Elish and boyd (2018) do, that the 'magic' of AI is only a mystification of a technology that becomes part of 'hype' and 'fear' cycles, which in the end disempower us in what we think we can do with AI. They therefore also argue that these cycles may only be countered by developing a rich methodological framework for data analysis referring to the very design process of AI. One may also extend this argument to a data ethics approach to the development and adoption of AI in society.

However, to develop a methodological framework to do this, we need a conception of AI as a digital data process that can be designed and governed by humans. That is, narrowing our focus on AI as designed data systems and data processes makes AI more manageable than governing a rogue independent agent in society. AI's gradual practical implementation in society has progressed from rule-based expert systems encoded with the knowledge of human experts, applied in primarily human and physical environments to systems evolving and learning from big data in digital environments with increasingly autonomous decision-making agency and capabilities. It is also this latter practical application of AI as digital data processing that I use.

Expert Systems

The history of the technological development of AI consists of social peaks and lows predominantly due to its various levels of practical commercial application and philosophically challenging ambitions. In his account of the history of the development of AI from the 1950s to the 1990s, the AI researcher and entrepreneur Daniel Crevier (1993) describes the endeavour to artificially construct intelligence as a striving to also uncover the complex essence of human thought. This was not a modest ambition, and AI research in the 1950s and 1960s was first and foremost experimental, performed inside research labs

with various aspirations to imitate human decision-making and thought processes in maths and computer processing. Consequently, in the mid 1970s the
AI research field experienced its first 'AI winter', where these original grander
ambitions lost traction in funding and investment communities due to a lack of
practical application (Crevier, 1993).

However, in the 1970s, the development of logic-based programmed 'expert
systems' created a new space for an initial commercial adoption of AI. As
a consequence, in the 1980s expert systems were created to support or even
replace decision-making in professional settings, where information was
collected from human experts and then coded as rules and procedures of the
computer (Alpaydin, 2016).

The promises of these systems to reduce the costs of human resources
were at first very grand. Crevier provides various examples from industries
in the 1980s that replaced human experts with expert systems with the aim of
reducing the cost of training people and moving field experts around to share
their knowledge and, for example, troubleshoot problems. One example was
the North American General Electric Company, whose experienced engineer,
David Smith, was the only person who could handle electric locomotive repair
problems, and who would therefore need to be physically transported around
to fix broken engines. In 1981, when Smith was considering retirement, the
General Electric company managed to codify his expertise into an expert
system named the Diesel Electric Locomotive Troubleshooting Aid (DELTA).
It contained hundreds of rules for troubleshooting and help, representing
Smith's knowledge. By 1984, DELTA could diagnose 80% of the breakdowns
and provide detailed instructions for performing repairs on broken engines
(Crevier, 1993, p. 198).

The expert systems of the early 1980s were promising in their prospects to
reduce costs and distribute and sustain expertise within a company. However,
many also soon proved to be less valuable, working only in limited settings
and with unsatisfactory results (Alpaydin, 2016). In DELTA's case, users were
supposed to take over the maintenance of the system after its initial development, but no one wanted to take on this responsibility and it was therefore
never used (Gill, 1995, p. 66). Some of the problems with the early expert
systems were caused by the development of technical environments, such
as an expert system being misaligned with a company's general computing
environment (Gill, 1995, p. 64). However, other problems with the systems
could be traced back to their inability to adapt to human environments, as was
the case with DELTA; for example, concerns about the liability of developers
and companies using the systems, problems solved by the systems not being
considered critical by users, users' resistance to externally developed systems,
or the loss of key developers of the systems (Gill, 1995).

Machine Learning Decision-Making Systems

What happened in the years following the creation and application of the first expert systems is in many ways also a story about the development of an increasingly digitalised big data environment, which enabled what is also referred to as 'machine learning' systems. Machine learning was the most practical application of AI in the 2010s. With machine learning, an AI system's knowledge essentially no longer has to be provided by human experts as the system will learn and evolve with data; accordingly, the system gains its autonomy and agency. Here, it is not the human expert that capacitates AI but rather digitalised data sets; a machine learning system learns with and further evolves on the basis of automated data processing.

David Lehr and Paul Ohm (2017) describe the data analytical capabilities of a machine learning process as 'an automated process of discovering correlations (sometimes alternatively referred to as relationships or patterns) between variables in a data set, often to make predictions or estimates of some outcome' (Lehr & Ohm, 2017, p. 671). One example is Apple's Siri, which analyses verbal questions and orders either directly on the device or by searching the internet. What makes Siri intelligent is that the program evolves and learns from data one provides through the questions one asks. In this way, the program assimilates itself by creating a data profile of the user, eventually becoming more of a personal assistant.

In the late 2010s, machine learning systems for analysing and acting on data in real time were increasingly embedded in data systems in all societal sectors and spheres, from healthcare to social networking. Professor of computer engineering Ethem Alpaydin (2016) credits this revolution of machine learning systems to the creation of the digital environment. In the 1980s, the invention of the microprocessor initiated the massive development of personal computers; as a result, computers, and later personal devices, were distributed widely in populations. The digitalisation processes of the 1990s going into the 2000s further enabled a pervasive immense collection of big data. Now all information, from colours in a photo to tones in an audio recording, could be transformed into a set of numbers and processed by computers (Alpaydin, 2016).

These technological developments paved the way for the fast-paced advancement of machine learning and a growing portfolio of internet-connected things, further facilitating increasingly autonomous behaviour and analytical agency of AI systems that learned and evolved via big data. To illustrate this, consider the CogniToy Dinosaur – a toy that used one of the most powerful machine learning models in the world, the Jeopardy-winning IBM Watson computer, to assess a child's interaction with it. The toy was not programmed with predetermined responses, but rather learned from a child's questions and responses and tailored its own responses to them. A child would say 'my favourite colour is red' and the dinosaur would respond 'okay, I will try to remember that', while storing this information for future more personalised play.

Notably, while machine learning cut out the human expert in some aspects, it did not entirely exclude human involvement in the very data design of machine learning systems. On the contrary, the degree of autonomy in an AI system based on machine learning depends on the human involvement in data processing, from problem definition, collection of data and data cleaning to the training of the machine learning algorithm (Lehr & Ohm, 2017).

Lehr and Ohm (2017) refer to this human involvement in machine learning processes as 'playing with data'. They argue that legal scholars have been too focused on the autonomy of machine learning systems by primarily concerning themselves with the 'running model' of the systems (the way they are adopted and used), while neglecting the data-processing activities that shape a machine learning system. Machine learning systems, as they state, are not magical black boxes with mysterious inner workings. In fact, they are the 'complicated outputs of intense human labor – labor from data scientists, statisticians, analysts, and computer programmers' (Lehr & Ohm, 2017, p. 717). In this perspective, we may consider the most widely used form of AI systems in the beginning of the 21st century – based on machine learning – as essentially a data processing practice with a degree of autonomy that may be influenced by different forms of human involvement, not only in the technical design but also in the organisation and application in society.

It is also this conception of AI that I use here. Rather than a scientific or technical term, I consider the concept's revival and application in a specific moment in history and do not ponder further on the potential 'intelligence' of AI (technologically or philosophically). I use the term in this way to address the more generic use of it in public discourse and, more specifically, in the European policy discourse of the late 2010s, which was concerned with the technological and social evolution of big data systems in society. I also want to argue that while the term AI was indeed the term most commonly used in this period, artificial human-level intelligence, as such, was in fact not the emphasis.

One of the first tasks that the EU's AI High-Level Expert Group (HLEG) set out to accomplish was to create a definition of AI. This technical definition, which was later published as an official deliverable of the group, emphasised the data processes and human involvement in AI:

> Artificial intelligence (AI) systems are software (and possibly also hardware) systems designed by humans that, given a complex goal, act in the physical or digital dimension by perceiving their environment through data acquisition, interpreting the collected structured or unstructured data, reasoning on the knowledge, or

processing the information derived from this data and deciding the best action(s) to take to achieve the given goal. (HLEG C, 2019, p. 6)

The European Commission policy and investment strategies that followed were also created to specifically secure the data resources of AI (for example, as expressed in the EU's data strategy published February 2020, European Commission H, 2020) or to ensure the development and access to big data AI-enhanced tools (products and services) to support the European private and public sectors (for example, in the EU's AI white paper published together with the data strategy in February 2020, European Commission I, 2020).

Artificial Intelligence in Society

Now, we have a technical definition of the predominantly in use and commercially available form of AI at the outset of the 21st century as a complex data processing system with a level of autonomy shaped by human involvement. AI enhances the analytical and sensory technical capabilities of big data systems distributed in networks via a myriad of digitally connected devices and things. However, these new technical capabilities are also a societal evolution as they are embedded in the very sociotechnical infrastructure of societies, thereby transforming sectors and the private and public spheres of citizens' lives in profound ways. Crucially, in the 2010s, the private and public sectors saw a fast-paced adoption of AI systems. AI applications were created for education, the environment, energy, healthcare, policy, financial IT, smart cities, mobility and sustainability, among other areas (Allam & Dhunny, 2019). If not implemented, strategies for their adoption in different sectors were developed with varying degrees of human involvement. The following are some of the examples from different societal sectors:

• The public sector

In various European countries, strategies were created to, for instance, integrate AI in public institutions to develop personalised assistance, chatbots and conversational platforms, as well as for socially scoring families and tracing vulnerable children. Moreover, the public sector saw the use of applications for automating civil servants' tasks, predictive policing and fraud detection (Spielkamp, 2019, AlgorithmWatch, 2020).

• The financial sector

In the financial sector, finance was transforming into 'cyborg finance' in which humans and machines 'share power' (Lin, 2014). For example, 80% of transactions on the Forex market (where the world's currencies are traded) were performed by robots (Bigiotti & Navarra, 2019). Finance 'Robo advisors' provided financial advice for investment management (Lieber, 11 April 2014) and financial institutions used AI applications for market analysis and to assess credit quality and price loan contracts (Financial Stability Board, 2017). Moreover, AI systems were used to assess and credit-score consumers (Pasquale, 2013).

- Social networking

The most common social networking services, provided by platforms such as Facebook and Google, were using AI systems to provide personalised ranking and recommendations, analyse the words and phrases of search queries and decide what other words have the same meaning, provide facial recognition to track users in photos, understand and respond to conversations, or detect misinformation and illegal and harmful content.[1]

- Smart cities

Cities were transformed with internet-connected things with sensors that collected data in real time. AI was integrated into city management and into engineering and construction to analyse distribution in real time and centralise data from various urban components (Allam & Dhunny, 2019).

- Health care

In health care, AI was used to support decision-making in the probability and estimation of diseases, personalised medicine, illness monitoring and treatment planning, critical care, diagnosis, treatment decisions and triage (WHO, 2018).

2.   AI ETHICS

In Norse mythology, Odin, the king of Asgård, the world of the gods, sits on his throne, long-bearded, cloaked, with his heavy helmet and sword, waiting to receive the dead Viking warriors from the battlefields of the human world. He has only one functioning eye: the other he sacrificed to gain an unconceivable amount of knowledge. Instead, he has his two ravens, Hugin ('thought') and Munin ('memory') (Orchard, 1997). They see and hear everything, they can talk, remember all, and predict the future. He depends on them, notwithstanding he lets them roam wild to scout the world for him. It is a trade-off, a delegation of his powers that he has to accept to be able to control the present and see the future, and so he also frets: *'Hugin and Munin fly each day over the spacious earth. I fear for Hugin, that he come not back, yet more anxious am I for Munin'* ('Grímnismál', Thorpe, 1907). These concerns of an ancient Viking god spell out a human anxiety about loss of agency that also applies to the very present debate on the ethics of the powers of AI. Which trade-offs are we willing to accept in our yearning to surpass the limits of the human body and mind when developing, adopting and regulating AISTIs? Here, we might learn from Odin's anxiety about the potential loss of memory (Munin); because what is a thought (Hugin), an intellect, without the situated dynamic qualities of a human memory (Munin) and experience?

AI had in the early 21st century transformed from being a mid-20th century scientific endeavour and sci-fi curiosity into a 'young' sociotechnical system with rapid societal adoption. As described, the new technical AI capabilities of big data systems were also a societal evolution. The systems of the infrastructures of private and public sectors were gradually transforming: they were becoming less physical and human-controlled, and increasingly based on digital big data and enhanced through AI. As follows, decision-making processes in these sectors were increasingly also informed by, and even being replaced by, big data AI systems of prediction and various types of risk or potential analysis.

What did this mean in practice? Let us examine a few examples. Recommendation and personalisation systems profile and analyse people's personal data and shape for them what they see and read and who they engage with online. Systems for autonomous driving scan the street in front of the driver, evaluate the risks and worth of the different objects in the way, and decide who or what the car hits in case of an unavoidable accident. Judicial risk assessment systems look for patterns in backgrounds of defendants to inform a judge about who is most likely to commit a crime in the future. Triage systems process the medical and demographic history of patients to decide who gets the kidney. All of these processes of partly or primarily autonomous decision-making AI systems comprise ethical dilemmas that are increasingly extended into AI systems. By way of illustration, on a macro-level, as citizens in a democratic society, what exact degree of choice and insight into the political processes being facilitated and transformed by AISTIs should we have? Or, on a micro-level, who should the car hit? The young person with a criminal background or the elderly person who never committed a crime?

As AI systems were envisioned, adopted and embedded in societal infrastructures in the 2010s, their ethical implications were also materialising in the shape of moral decisions and choices intertwined with the complex data processing of AI systems. Hence, concurrently with a renewed public focus on AI, 'AI ethics' emerged as a research field concerned with the ethical implications of AI systems. Although various terms are used to describe the different aspects of this field, I here use the term 'AI ethics' to describe the general research field that addresses areas of concern in regard to the ethical implications of the practical application of AI in society.

I will now examine how the ethical concerns I voiced earlier – regarding different levels of human involvement in AI systems' design, adoption and consolidation in society – can also be traced in an overarching general theme within the 'AI ethics' research field. Furthermore, I associate these concerns with the aforementioned imagined scenarios regarding AI's threat to humanity and human control or the potential of autonomous AI to surpass human deficits. As I will illustrate, due to the varying aspirations and conceptions of AI, the practical application of AI ethics also deals with very different levels of human involvement in the design and governance of AI. In my view, the most valuable applied AI ethics approach is the one that prioritises the highest level of human involvement in AI development.

The ethical implications of AI systems' role in human and societal decision-making processes are also a general theme in the AI ethics debate. It spans, as I will show in the following sections, from discussions regarding the role of machine agency in the moral world of humans as either a positive or destructive transformative force, to applied AI ethics' methodologies and frameworks. The latter, in particular, can be examined from the point of view of embedding different degrees of human agency and involvement in the very design and organisation of AI decision-making processes.

While I consider a perspective on the very design of AI a constructive and highly relevant contribution, I also propose that it is only one of the applied ethics components of an ethical governance framework for AISTIs. Thus, I am not suggesting here that a data ethics of power is an 'ethics by design' solution. While I will, later in the chapter, illustrate how we can trace interests in data in the very data design of AI, I do not offer a design solution. Rather, the contribution of a data ethics of power to the AI ethics research field is, I argue, a targeted reflection on human power in AI development; that is, the varying degrees of human involvement that we assume and request in the development and adoption of AI. Furthermore, to narrow down the discussion on AI ethics to a data ethics of power, I will primarily concentrate on the levels of autonomy and human involvement in AI systems' data processing. Here, I address not only the developer side of data design but also the social and cultural adoption and consolidation of AI data systems.

From Human-Dependent Systems to Autonomous Systems

AI systems have, in their short history of societal adoption, been used to support or replace human decision-making processes with various levels of autonomy. The agency and autonomous behaviour of AI systems were, in their practical application, not an objective per se, but a fundamental feature of the system's ability to adapt to real-life decision-making processes.

In the 1970s and 1980s, expert systems were created to aid humans in decision-making, such as when troubleshooting and guiding decisions on how to repair machines or diagnose infectious diseases (Crevier, 1993). They were composed of a 'knowledge base', which consisted of a range of facts and 'IF-THEN rules' based on the knowledge of human domain experts, and an 'inference engine' using logical inference rules to deduct new knowledge (Alpaydin, 2016, p. 50). Previously, I described these early expert systems' inability to adapt to human environments as a key reason they did not succeed in societal adoption. This also included the way in which they represented the real world. The logical rules of the systems were simply too rigid to represent it, they could not represent the nuances and gradations of life. Alpaydin uses the example of age; one is not just 'old', but we are growing old gradually, and this process cannot be captured by a figure (Alpaydin, 2016, p. 51).[2] The early expert systems' evidence (via pre-programmed knowledge and logical rules) was evidently very limited and often also faulty in terms of the representation of the nuances of a real environment. They were failing in presenting valuable decisions and their practical application was, for this reason, very limited.

Contemporary AI systems constitute improvements of these original expert systems in decision-making. However, increasingly they can reason, make decisions and learn by themselves via complex multi-layered data processing and sensors which make them capable of perceiving complex environments (HLEG C, 2019, p. 3). They are therefore also better (but of course never perfect) at analysing the nuances of real-world settings (Alpaydin, 2016, p. 52). Machine learning which uses algorithms based on the concept of neural networks, such as in deep learning, dynamically uses input data from sensors, which is then processed progressively in such a way that each layer of analysis takes its input from the previous one to produce a decision (Alpaydin, 2016, p. 85).

Moral Machines

Now, it is one thing to build evidence for the troubleshooting of a faulty engine and make a decision regarding its repair; it is another thing altogether to build evidence with the nuances required to make complex moral decisions that affect human lives. Increasingly, AI systems are implemented in settings that involve ethical reflection and moral decision-making per se, or that are transformed by the systems in ways that produce new ethical implications.

Most famous is the autonomous vehicle ethical dilemma presented by Awad et al. (2018) in their 'Moral Machine' experiment, which explored the moral decisions involved in driving a car that is in an accident involving pedestrians. By way of illustration, if the only choice is between causing the car to crash, with yourself, as the driver, hitting two elderly people or hitting one young person, what would be the morally correct decision to make? Awad et al. developed a serious online game with ethical dilemmas associated with scenarios such as this to examine people's moral choices and to create, as they said, 'a global conversation to express our preferences to the companies that will design moral algorithms, and to the policymakers that will regulate them' (Awad et al., 2018, p. 63). They focused on the 'running model' (Lehr & Ohm, 2017) of the AI system's decision-making process; that is, the moment it is deployed and makes a decision entirely on its own without human involvement. In this way, they also imagined an everyday life in the future where machines will replace human decision-making and act autonomously:

> Never in the history of humanity have we allowed a machine to autonomously decide who should live and who should die, in a fraction of a second, without real-time supervision. We are going to cross that bridge any time now, and it will not happen in a distant theatre of military operations; it will happen in that most mundane aspect of our lives, everyday transportation. (Awad et al., 2018, p. 63)

The 'Moral Machine' experiment reimagines the famous trolley problem, where different scenarios and ethical dilemmas are tested. It has become the most used example of AI ethics; however, I do not consider it the best one in the context of understanding the practical implementation of a 'human approach' to AI. In fact, the very programmed choice of the machine is not the ethical dilemma we want to consider first. I want to start before that: to think about the ethics of a machine that does not make autonomous decisions without us. What we want to ask is this: how do we want the machine to help us make the decisions we, as humans, want to make? How does the machine complement a human environment? In this way we could think of a future alternative to the one envisioned in the 'Moral Machine' experiment in which no critical decisions can be made without human involvement. In European law (General Data Protection Regulation [EU] 2016/679) there are, for example, provisions that prohibit decisions regarding individuals based solely on automated data processing without human involvement if they significantly affect an individual (Regulation [EU] 2016/679, Article 22).

Ethical Implications of AI Decision-Making

In assessing and mapping the ethics debate on 'algorithms', Mittelstadt et al. (2016) identify various types of ethical concern connected with the way in which algorithms process and make correlations in data (make evidence out of data) to reach decisions. The ethical implications, such as discriminatory decisions, can be immediately 'visible' outcomes – actions that can be discerned as 'unfair' in the moment of observation. However, ethical implications may also be societally transformative in ways that are not observably harmful in the moment of implementation (ibid., p. 5). Here, the challenges to autonomy are an overarching concern that considers most predominant personalisation algorithms and the construction of 'new choice architectures' which may nudge our behaviour and control our decisions to varying degrees (Mittelstadt et al., 2016, p. 9). Concern about the challenges to the way in which we conceive of and deal with informational privacy – brought about by big data collection and processing in the form of profiling algorithms – is another example. Lastly, they consider a horizontal concern regarding the traceability of algorithms (Mittelstadt et al., 2016, p. 12). The very design of the complex data processing of algorithms' evidence is often difficult to trace and therefore generally complicates the identification of responsibility for the ethical implications of algorithms. However, the level of human oversight of increasingly complex systems may also be complicated by other factors. A lack of education or other human-level capacities and factors (such as one's level of awareness, ethical reflection, and other cultural and social factors) may equally make it more difficult for humans to identify and/or correct design, which also carries ethical implications. Lack of human oversight also means that control and agency are gradually moved into the system.

Machine Ethics

The AI transformation of human–machine relations raises many ethical concerns in regard to human agency and involvement in systems with increasingly distributed moral decision-making. 'Machine ethics' (Anderson & Anderson, 2011) focuses on the ethical behaviour of autonomous AI agents. A foundational prediction here is that in the future, human involvement will be minimal, and therefore machines must be equipped with ethical and moral capabilities. As follows, we need to develop theories and methodologies to train machines to act ethically:

> Theoretically, machine ethics is concerned with giving machines ethical principles or a procedure for discovering a way to resolve the ethical dilemmas they might encounter, enabling them to function in an ethically responsible manner through their own ethical decision making. (Anderson & Anderson, 2011, p. 1)

One strand of the machine ethics research field even considers AI agents with ethical capabilities as a way to improve human moral decision-making. Intelligent machines are here perceived as morally superior to humans (Anderson, 2011; Dietrich, 2011). They can help us create universal ethical principles by surpassing the relativism of a human moral decision-making driven by a self-serving interest (Anderson, 2011). Following this, Seville and

Field (2011) envision an 'ethical decision agent' that can help people make ethical decisions by pointing to consequences of their decisions or a virtual reality to create 'ethical experiences'. This agent, they argue, would be more impartial and add consistency to moral decision-making.

The literature offering perspectives on machine ethics provides valuable insights in regard to the design and implementation of technically mediated components of moral decision-making, which are increasingly distributed in external technological systems. The recognition of a new type of technological agent that actively participates in moral decision-making processes, and consequently also in shaping our ethical experiences, is of particular importance here. However, as I argue in the following section, AI agents do not take on moral responsibility as more impartial moral agents imposing universal moral norms. Moral, or rather 'ethical' responsibility, I maintain, will always be intrinsically part of human involvement in the design, development and implementation of AI. This also implies that responsibility for any moral action and ethical implication of an AI system can and should only be that of the humans involved (Bryson, 2018). As such, the very foundation of an argument for building artificially moral agents with superior moral skills to correct the errors of human moral reasoning is also challenged (van Wynsberghe & Robbins, 2019).

AI Moral Agency and Human Ethical Responsibility

In 2017, the European Parliament adopted a resolution with recommendations on Civil Law Rules on Robotics. In this resolution, the question of legal liability for harmful actions and implications of autonomous systems was raised. 'Consider', the resolution therefore states, a 'new legal status' for autonomous robots, possibly an 'electronic personality' (European Parliament, 16 February 2017). I believe that even *considering* the responsibility of AI agents, as such, has dire ethical implications, as it implies that we also accept autonomous ethical agency. We do not need to do this because human involvement and agency is, although at times difficult to discern, always present in AI. It is present in the *how* of the data of AI, as I have illustrated in the previous sections; in the laws that frame AI; in the technological cultures of its design; in the way in which we handle and adopt AI in society; and, crucially, human agency is present in how 'AI autonomy' is socially perceived, accepted or rejected. Thus, while AI does indeed have technical decision-making capabilities and may be imagined in the context of autonomous machine agency, human involvement always plays a role. Accordingly, what has to be done is to enhance and support this 'human factor' in the design, development and adoption, as well as the legal frameworks, for AI.

In the 'AI ethics' debate, two opposite poles represent the threat to human moral (ethical) agency and control, or the potential of powerful and humanly superior machines for human moral (ethical) improvement.[3] However, there is a middle way. By accepting that AI systems are moral agents, I maintain that we do not simultaneously acknowledge that they are also ethical agents; that is, *ethically responsible* agents. All that is recognised is that humans are not the *only* agents that shape the moral architectures of the environments in which we live (Adam, 2008). Said in other words, AI does not cause a 'responsibility gap' where blame for ethical implications has no object (Tigard, 2020). There is always an ethically responsible human dimension to AI. By way of example, we can think of the ethical implications of the application of an autonomous vehicle that kills a pedestrian in terms of the complexity of the human conditions that led to the accident; as the result of a network of human processes – from design choices to implementation; and even a result of the adequacy of the laws that frame the development and the use of the autonomous vehicle, as well as the rules and the shape of the streets on which it drives. This does not mean that AI systems escape legal accountability; all it means is that only humans can be ethically responsible.

Now, if we consider AI decision-making systems as components of socio-technical architectures of distributed moral agencies where humans and nonhuman agents are intertwined in shaping moral experiences, questions regarding the moral status of nonhuman agents become more practical than existential in nature. In this case, we do not even need to ask if machines should or can have human-level ethical agency or whether they can be ethically responsible, but rather we should attempt to find a way to ensure that humans continue to be involved in a meaningful and, crucially, responsible manner.

Here we may use Bruno Latour's description of the 'moral agency' of technological artefacts as delegated nonhuman actors that enforce human laws, values and ethics (Latour, 1992; Latour & Venn, 2002). He argues that technological artefacts are indeed 'strongly social and highly moral' (Latour, 1992, p. 152) and they work by the prescription of laws and orders that are 'inscribed or encoded in the machine' (Latour, 1992, p. 177). As Latour illustrates it, a technological artefact such as a seat belt can lock our bodies in positions we do not wish to be in. It is designed to do exactly this, it does indeed enforce the laws of car safety, and will certainly let us know with an insistent beeping sound if we are not ascribing to these laws.

However, technologies are not just passive expressions of moral intentions; they are what Latour refers to as 'technical mediators' (Latour & Venn, 2002, p. 252). Moral intentions and actions are actively translated in the technical design intertwined with various possibilities that are in constant negotiation with use, laws, culture and the society in which we act. We can here use the example of 'word embedding' machine learning methods, which are used for the language processing of online search engines. Examining the most commonly used models, Bolukbasi et al. (2016) found that word clusters were created in which words such as architect, philosopher, financier and similar titles were grouped together semantically as 'extreme he' words, whereas words such as receptionist, housekeeper and nanny were grouped together as 'extreme she' words (Bolukbasi et al., 2016, p. 2). Hence, they conclude that the blind application of this model could contribute to discrimination in society. The work of a developer of this type of machine learning model clearly constitutes a moral action that entails the delegation of moral values and choices to a technological moral agent that then shapes the moral factors of a social environment. As Bowker and Star (2000) argue, classifications of information are never neutral. There is always a 'moral' dimension (Bowker & Star, 2000, p. 5) to the work of a developer of an information system, such as the word-embedding model of a search engine. Thus, this human actor has a crucial role as the actor who inscribes the programming language of the technology's delegated moral agency (Latour, 1992). However, the translation of this type of technological moral agency into moral action and/or implications is not a straightforward process. If we trace the moral agency that in this example results in discrimination, several active actions are not solely those of the human developer of the machine learning model; neither are they solely those of the 'machine'. The machine learning model (nonhuman actor) actively amplifies existing human bias (human actor) in its training data; that is, it is learning from and evolving (creating gender-biased word clusters) from Google news articles. Nevertheless, it is also 'blindly' developed, accepted and enacted in society as an objective representation of information by human actors (the user and the developer).

We need to examine the relations between the distributed moral agency of active human and nonhuman actors. Social and ethical implications are in this perspective not just the result of a human intention; neither are they of the design and actions of nonhuman actors. Rather, they are consequences of a network of actions and competences distributed between these different agents.

What does this model of distributed moral agency then mean in terms of applied 'AI ethics'? How do we consider the ethical implications and actively apply and test ethical considerations in the development of AI systems? Above all, it means that we cannot just design a moral norm into a technology and thus produce a 'Moral Machine'; we need to address the way in which ethical implications evolve in environments of distributed moral agencies between human and nonhuman actors.

When describing how to design 'artificial morality' in artificial agents, Allen et al. (2005) offer two approaches. One is the top-down approach, in which the machine is designed to act according to specific moral principles;

that is, moral theories may be used as the programmed rules for the selection of ethically appropriate actions (Allen et al., 2005, p. 149). We return again to the 'Moral Machine' experiment, which has the objective of creating a global foundation for designing an autonomous machine that may take moral action by itself based on the moral norms inscribed by humans. This model deals with the extension of the moral intentions of a human global society into external technological systems. However, this model does not take into account the type of dynamic distributed moral agency that I have just described. This is where we may use Allen et al.'s (2005) second approach, where we do not impose a specific moral theory but aim to provide environments (with, for example, meaningful human involvement and agency) for AI agents in which appropriate behaviour is selected and rewarded (Allen et al., 2005, p. 151). In this way, the machine acts ethically by dynamically evolving in ethical human environments. This description of a critical applied ethics approach to AI leads to the following discussion of the contexts of power relations and interests in which AI technologies evolve.

Interests and Power Relations

The AI ethics research field comprises a recent concern with the ethical impli-cations of increasingly autonomous data systems and algorithms. However, the discussion merges with previous debates regarding the 'neutrality' of computer technologies. The conceptualisation of the entrenched values of a computer technology design was originally formulated by information science scholar Batya Friedman et al. in the 1990s, and has since then been further explored in the value-sensitive design (VSD) framework (Friedman & Nissenbaum, 1995, 1996, 1997; Friedman, 1996; Friedman et al., 2006; Flanagan et al., 2008; Umbrello, 2019, 2020; Umbrello & Yampolskiy, 2020). In this perspective, a computer technology is never neutral, but rather embodies moral values and norms in its very design (Flanagan et al., 2008).

In VSD the embedded values of a technology are addressed as ethical dilem-mas or moral problems to solve in the very design and practical application of computer technologies. The ethical implications of a computer technology can therefore be analysed by examining the technical design, which can be designed in 'ethical' or 'ethically problematic' ways. Together with another information science scholar, Helen Nissenbaum, Batya Friedman (1996), for instance, illustrated different types of bias embedded in existing computer systems used for tasks such as flight reservations or the assignment of medical graduates to their first employment, and presented a framework for addressing this in the design of computer systems.

The VSD approach has been similarly employed in studies of the values embedded in AI system design, extending the analysis to the entire lifecycle of AI (Friedman & Hendry, 2019; Umbrello & Yampolskiy, 2020). The data systems and mathematically designed algorithms of AI are not impartial or objective representations of the world. Consequently, they also trigger actions and societal effects (decisions and suggestions) that are not 'ethically neutral' (Mittelstadt et al., 2016, p. 4).

One strand of research of the late 2010s specifically addressed the 'non-neutrality' of AI systems with specific reference to the power dynamics of the adoption and implementation of AI and big data systems in the public and private sectors of the period. Several concrete examples and case studies of the social and ethical implications of AI systems were used to highlight the social power relations and interests at play in the development and societal adoption of AI.

Cathy O'Neil (2016), for example, described the darker side of big data systems, or what she referred to as 'Weapons of Math Destruction' (WMDs), which were implemented in US educational and public employment systems for credit scoring and insurance assessments. Her primary concern was that these big data systems were being deployed without private and state actors questioning the assessment of their social implications as neutral and objective systems capable of replacing human decision-making and assessment. She illustrated the at-times devastating consequences for citizens. For example, teachers were fired based on rigid machine-based performance assessments that did not consider social contexts and human factors (O'Neil, 2016, p. 5), and furthermore, people from less desirable demographics received lower credit scores based on their computer's location (O'Neil, 2016, p. 144).

Frank Pasquale (2015) was equally concerned with the use of automated processes to assess risks and allocate opportunities. He illustrated how complex algorithms were developed and deployed to sustain a 'Black Box Society' where the data processes of algorithms were protected intentionally as trade secrets to sustain the information monopolies of powerful industries. These industry interests also authorised computers to make decisions without human intervention.

Another famous case study of the power relations at play in the adoption of an AI system was the 'Machine Bias' study published by the news site Probublica (Angwin et al., 2016). Here, the investigative journalist Julia Angwin, together with a team of journalists and data scientists, examined the private company Northpoint's COMPAS algorithm, which was used to perform risk assessments of defendants in the US judicial system and assess the likelihood of recidivism after release. They found a bias in the algorithm that had the tendency to designate black defendants as possible reoffenders twice as often as it did white defendants. Moreover, it classified white defend-ants as a low risk more often than black defendants.

Safiya Umoja Noble (2018) described the power of the search architectures of Google's algorithms as oppressive ('algorithms of oppression'). As she illustrated in her investigations, they not only reinforce bias in society when they replicate prejudices against, for example, African American girls in sexualised and discriminating search results. They also enact bias when presented and received as natural objective categories of reality.

To examine and reveal the potential discriminatory agency of an algorithm is essential in a society where ethical implications arise from the moral agency distributed among human agents (for example, developers or judges who use a risk-assessment tool) and nonhuman agents (the tool's data design). We can even offer critical applied ethics methodologies for mitigating bias within the very data processing of the machine learning model of the tool. Lehr and Ohm (2017), for instance, have presented several ways of intervening in the 'playing with data' stages of a discriminatory machine learning model; for example, by examining the way in which it facilitates the translation of disparities in training data into prediction disparities, generating less accurate predictive rules for minority groups than for others (Lehr & Ohm, 2017, p. 704). However, this can only be done if we have the option to trace the actual data processes of the algorithm. Mittelstadt et al. (2016) argue that the traceability of algorithms is complicated by the increasingly complex data processing of algorithms and humans' capacities (such as awareness and education) to identify and/or correct the data design of an algorithm. However, there is also another angle on this. Angwin and her co-investigators (2016) had to provide their evidence for the 'Machine Bias' study without accessing the 'playing with data' stage (Lehr & Ohm, 2017) of the COMPAS software, as its algorithm was protected as the proprietary property of the private company behind Northpoint. Their critique of the system was complicated by the commercial interests of the private company behind it. Thus, they could only study the 'running model' (Lehr & Ohm, 2017) by comparing different data sets with public records requests (Larson et al., 2016). This very inaccessibility and 'autonomy' of a sociotechnical tool, enforced by humans with the interests of a company, are ethical implications in themselves. It is here not just a question of technical complexity and human education and capacity, but also the result of societal power relations expressed in legal protections or freedoms. The lack of traceability can therefore not be reduced to a clash between an autonomous, technically complex moral agent and the individual human's capacities. As Pasquale argues, and the 'Machine Bias' case study illustrates, the general social power relations and interests at play also complicate audits and interventions by preventing access to the very data design of a running model.

Data Ethical Governance in AI Policy

What role can 'data ethical governance' play for policymakers that will help them address AI? As argued in the first part of this book, and further expanded on in this chapter, we need to address the complexity of a sociotechnical system in order to guide it. 'AI ethics' is essentially an applied approach to an AI system's design. This will take us part of the way. However, changing only the design of AI will not change the direction of a sociotechnical development. What we need to do is to address the distributed moral agency of AI in the contexts of the powers and interests in human societies.

In this chapter, I have discussed the autonomy of AI systems as an ethical problem to solve. Subsequently, I have examined the level of human involvement in AI development and adoption as one key component to consider with an applied 'AI ethics' approach. I have also focused specifically on the data processing of AI. I have examined the level of autonomy and thus human involvement in AI as something that can be addressed in the micro-contexts of developing AI. However, and crucially, I have also examined the autonomy of AI systems in the context of human societal power dynamics on a macro-level; that is, the level of AI autonomy and human empowerment as something that is shaped by different interests in society. Following this, we can now think of some examples of 'data ethical governance' that may help shape public policy proposals and activities that specifically address human power and the level of autonomy of AI:

– **Legal Frameworks that 'Defend Human Powers'**

In his book, *The Black Box Society* (2015), Pasquale proposed that legal frameworks are created for what he described as an 'intelligible society' in which decision-making processes are always intelligible to all humans involved on a technical, organisational as well as a societal level. This will require what he refers to as 'humanising processes' (Pasquale, 2015, p. 198); that is, the establishment of company and policymaking practices that embed 'human judgement' and involvement in automated decision-making processes (Pasquale, 2015, p. 197). Pasquale provided some implementable examples, such as human experts informing policymakers when their understanding of a technology is incomplete (Pasquale, 2015, p. 197). However, as he later proposed and elaborated in his 2020 book on a set of new laws of robotics, what we really need is a general framework for defending 'human expertise in the age of AI' (Pasquale, 2020). We may think here of examples from concrete legislative frameworks developed in Europe in the 2010s, for instance, as I have mentioned earlier, the legal provision of the GDPR's Article 22 on automated individual decision-making and profiling (Regulation [EU]

2016/679), which basically aims to ensure the 'human factor' in such systems. In 2020, the European Commission furthermore published proposals for a Data Governance Act (DGA), a Digital Service Act (DSA) and a Digital Markets Act (DMA) with the overall objective to harness the power of large online platforms, or what was also referred to as 'gate keepers'.[4] The Data Governance Act, for example, covered a call for investing in and supporting the development of trusted 'data intermediaries' (that is, data trusts and stewardship models) to balance data asymmetries between major big data online platforms and individuals. The DSA, on the other hand, emphasised safeguards in regard to automated content moderation and data access to enable external audits and risk assessment of large online platforms' AI systems.

– Bottom-up Governance Approaches Shaping Human Involvement in AI Systems

Regulatory frameworks such as the GDPR and the proposed (in 2020) DGA, DSA and DMA work as top-down requirements on the development and adoption of AI. However, other bottom-up governance approaches may also shape human involvement in AI development and adoption. By way of example, public institutions can use public procurement strategically to encourage the development of AI systems with heightened levels of human involvement (Hasselbalch, Olsen & Tranberg, 2020). Governmental and intergovernmental investment schemes can support start-ups that differentiate themselves on the global market with ethics-by-design innovation, products and services. In the micro-context of design and development, engineers and developers need tools and methods to include human involvement components in their work with algorithms and machine learning models, such as 'human in the loop' methodologies (Zanzotto, 2019), techniques for explainability and traceability (Gilpin et al., 2018), anonymisation techniques (Augusto et al., 2019), verification and validation and risk assessment tools (Menzies & Pecheur, 2005). Here, educational programmes can be implemented to increase developer competences and awareness, and shared technical engineering standards, can be supported.

In conclusion, as these examples illustrate, the 'data ethical governance' of AI in a policy context is not simply a matter of discovering and harnessing the moral agency and ethical implications in the very design of AI. It means encompassing the entire chain of the distributed moral agencies of human and nonhuman actors.

Data Ethical Implications of AI

A data ethics of power suitable for AISTI governance must encompass the special power dynamics of AISTIs. To conclude this chapter, let us consider the special powers and data ethical implications of AISTIs. The computer hardware and software of the AI systems that store and handle big data are embedded in society and, like bridges, streets, parks, railroads and airways, they form our spatial environment, although they are different from traditional infrastructures. Roads and bridges, for instance, form the basic material architectonics of society and thus provide or limit access to places, but they are passive, so to speak, when mapping and expressing human motives, morals and social laws. AISTIs transform the very objective material qualities of space. Quite literally they transform space into interconnected digital data. GeoAI (Krzysztof et al., 2020), for example, is a term used to describe the integration of AI systems and geography based on the analysis of data that contains georeferenced information ('GPS trajectories, remote sensing images, location-based social media, spatial footprints of buildings, roads, and parcels, global elevation data, land use and land cover data, population distribution, and so forth'; Hu et al., 2019, p. 2).

However, AISTIs are not just digital data extensions of material space. Recalling Lapenta's (2011) depiction of the 21st century 'Geomedia' (Lapenta, 2011), AISTIs also lock us in specific positions, providing or denying access based on the processing of personal data. They are mediating spaces that merge the human body, social and individual experiences, physical space and location into interoperable digital data, blurring their lines of separation when integrating them into the designed spatial architectures of a virtual infrastructure. In this way, AISTIs function as the 'new organisational and regulatory systems' articulating and organising social interactions (Lapenta, 2011, p. 21).

Geomedia constitute our everyday life spaces merged with mediated data spaces. These include navigation tools such as Google maps and other location-based services such as the ride-sharing services Uber and Lyft. Then, there is also the 'Geo-spatial intelligence' system, Sentient, which at the time of writing is under development by US intelligence programs. The idea is that it will work on satellite pictures of the world with time and location stamps integrating all data, and eventually it will enable instantaneous and omnipresent AI analysis and strategy development for the US military and intelligence services (Scoles, 31 July 2019).

AISTIs are active infrastructural practitioners. They sense, learn and act based on what they learn, and evolve autonomously or semi-autonomously based on their interconnected big data environments. With a component of autonomous decision-making and behaviour, they actively shape the space they occupy. Crucially, I wish to argue here that AISTIs therefore also actively participate in transforming the structure of our ethical experiences and critical practices. That is, they constitute an ethical experience, so to speak. Lefebvre (1974/1992) described the architectonics of a space as something that does not just 'exist' but is experienced. It is defined by a body's movement and the sensing of its borders and directions. He called this a 'bodily lived experience' (Lefebvre, 1974/1992, p. 40). That is, space is, as I also described in the first part of this book, material, social, but importantly also lived and experienced by humans.

Along these lines, we may also consider AISTIs to be actively self-producing spaces that amplify our experiences of a specific scientific, ideological and aesthetic paradigm. That is, as the historian Paul N. Edwards expresses it, they are modernity embodied in a lived reality of control and order (Edwards, 2002, p. 191). Think, for example, of Amazon's automated tracking and termination system, which it deployed in its warehouses in the 2010s. This was lived and experienced by Amazon workers, who were pressed to 'make rate' to pack hundreds of boxes per hour (Lecher, 25 April 2019): 'Amazon's system tracks the rates of each individual associate's productivity and automatically generates any warnings or terminations regarding quality or productivity'.[5] Or think of the experience of a student in the school district of Andhra Pradesh, India, where Microsoft's Azure Machine Learning was deployed to identify those at risk of dropping out of school. In 2018, the AI tool had, according to reports, identified 19,500 students at high risk based on predictive analyses of data such as gender, socioeconomic demographics, academic performance, school infrastructure and teacher skills (Surur, 22 April 2018).

To summarise, with big data we created a quantifiable, measurable space ready to act on. With AI we created an agent, the hands and the brains, of the big data infrastructure; however, it is a very particular kind of agent directed at managing future risks and potentials. Thus, AISTIs do not only produce space. They are 'Destiny Machines' (Hasselbalch, 2015) that also act on the temporal categories of individuals and societies by utilising the past (big data repositories) for the sole purpose of controlling and streamlining the present and future into something useful within the boundaries of the system design that is shaped by different interests in the data of the system.

In other words, AISTIs are spatial and temporal architectures interrelated via streams of big data. Their agency in the world is empowered by a seemingly random data interconnectedness that tells us where each of us needs to be in the larger scheme of things. But it is an agency without human intention. It has no interest in where each of us really wants to be or where we collectively ought to be. AISTIs provide us with agencies like the agency of the 'holistic' detective Dirk Gently, who says 'I may not have gone where I intended to go, but I think I have ended up where I needed to be'.[6] Dirk unscrambles mysteries in ways completely incomprehensible to any human, even himself, by accepting his position in the universe's interconnectedness of things that, time and again, take him on journeys across time and senseless places. He does so without questioning it, and without intention, because he knows that the universe will always place him exactly where he needs to be. By itself, this is AISTIs' core ethical challenge to human power and agency.

NOTES

1. Facebook and Google's AI sections: https://ai.facebook.com/, https://ai.google.
2. Here, we could use other examples from the more present world of social media content moderation, such as the meaning of concepts as for example 'indecency' and 'harmful'. Needless to say, time and again, the most publicly controversial cases of content blocking and take-down on social media portals in the 2010s stemmed from different interpretations of what constitutes 'harmful content', and critically, the social media platforms' automated content moderation systems' decision-making power and thus enforcement of specific politics and values in the public online sphere. See also the European Commission's original 1996 distinction between 'illegal' and 'harmful' content online and accordingly the different legal responsibilities in COM(96) 487 Final Brussels, 16.10.1996.
3. See my distinction between 'moral agency' and 'ethical agency' in Chapter 6 and the Terminology section for further elaboration.
4. See the European Commission's Digital Services Act package (December 2020) and Data Governance Act (November 2020).
5. Documents obtained from Verge https://cdn.vox-cdn.com/uploads/chorus_asset/file/16190209/amazon_terminations_documents.pdf
6. From Douglas Adams' *The Long Dark Tea-Time of the Soul* (1988).

4. Data interests and data cultures

'Tell me you work in tech without telling me you work in tech'

Tweet, 2021

While doing research for this book and participating actively in the 'data ethics' and 'trustworthy' AI policy and advocacy communities, I met many AI developers working on what they referred to as 'ethical AI' solutions. Working ethically with the data of AI, however, I also learned had many different meanings. There were developers developing AI for social purposes who would argue that not using all data was 'unethical' and there were the AI developers who, on the other hand, would argue for stricter privacy protections and accordingly for data minimisation in AI development. Even though they were, on the face of it, part of the same 'ethical AI' movement, they still did not have a shared conceptual framework for resolving conflicts between the data interests in the technologies they were developing. They were, I understood, part of different data cultures with different values and interests in data, organised in different – and sometimes clashing – conceptual maps of meaning.

All in all, when speaking with developers of AI technologies, I understood how various interests in data are constantly at play in the developmental phase of a data-intensive technology.

By way of illustration, as one of the AI developers I spoke to and who agreed to be quoted here explained to me: 'AI is very data hungry, so when we are building something, then we are thinking, what API is there that we can use for this, so what API Google, Bing or Amazon provides, like Bing or Amazon, what API is there that I can take the data, and make use of this…'

She worried about the data she was sharing with these cloud services: 'we just say that this is covered by their privacy policy, but we don't know exactly what are the clauses; I don't know, to be honest', and when in response to her concerns I queried her further about what she got out of using these platforms as an AI developer, she answered: 'I have good hardware, so I can run things way faster than I would on my computer, so I get the speed'.

In this way, a design choice was made with a trade-off between the data interests of users, the cloud service providers and the AI model. Later, she recalled a case in which the data interests of a group of people within a specific occupation trumped a business client's data interests. They wanted to use data about employees' performance for the AI system they were designing for the client. However, they had to negotiate with the trade union of the specific occupation that were protecting the interests of employees, because inferences could be made from this data about the performance of individual employees. When asked how a union got involved, she disclosed another design choice based on the negotiation and trade-off between data interests: 'we needed to talk to the union in order to use their data, but then in the end we didn't use it'.

Why does a data designer make the choices she does when considering different interests in the data design of the technology she is developing? What role does her cultural environment play? In this chapter, we will try to understand and explore the micro-settings in which the technical components of a sociotechnical infrastructure, such as AISTIs, take material form as expressions of interests orchestrated in cultural systems of meaning making. An understanding of this cultural organisation of meaning, I argue, is essential to the 'ethical governance' of the complexity of sociotechnical change, because shared cultural conceptual maps of meaning making are what enable us to act with shared ideas and purposes.

In the previous chapters, we looked at the shape of sociotechnical change as a complex of human and nonhuman factors that are thrown together in design, adoption, use and governance. However, as I continue to argue in this chapter, they are not thrown together in an arbitrary fashion. Sociotechnical change is made, it has politics, it has cultures, meaning that it is not neutral or 'natural', but embedded with interests. It is not even *it*, but, as also illustrated in the example of the 'ethical AI' design movement, it is *many*. Many embedded interests; many taken-for-granted cultures; many views of the world, priorities and conceptual frameworks, in harmony or in conflict. This also means that the 'technological momentum' (Hughes, 1983, 1987) that enables a consolidation of change represents a compromise between these multifaceted cultural interests.

According to Hughes, each developmental phase of a sociotechnical system's evolution produces a specific 'culture of technology', which he also defines as the environment of the sociotechnical system and the sum of invested interests (Hughes, 1983, 1987). We may consider the culture of a sociotechnical system a particular 'normality' (Kuhn, 1970), a knowledge foundation, worldview and conceptual framework for the practices of the developers designing its technical components, the lawmakers governing its adoption in society and the citizens in whose everyday lives it is incorporated into. Culture is what shapes the system's technological momentum. The fundamental idea here is that competing cultures, and accordingly competing interests, must convert to the dominant culture of the momentum or perish (Hughes, 1983). As such, technological change is first and foremost a negotiation between interests and a question of power and very human interests in power and dominion.

1. INTERESTS AND TECHNOLOGY

Large sociotechnical systems, such as AISTIs, transform and evolve in a multifaceted mix of relations. Each component of this development expresses its own history of knowledge, priorities, needs and available means. By way of illustration, personal AI assistants are software programs that interact with individuals answering questions and performing a variety of tasks for them. Based on various AI components, such as voice recognition and machine learning, they are designed to act on behalf of and respond to individual users in a personalised manner. In a way, one can also describe them as the individual's representative in the technical design of AISTIs. But in what way do they represent individuals? A data protection law may set the context of legal requirements for the assistant's data design; fields of scientific studies will frame its technical and social potential; a developer will design its pre-set goals and priorities; a company behind might request a data design fit for marketing purposes; the user of the AI assistant will shape its self-learning process with their personal data; and so on. All 'actors' arrive with pre-set requirements and priorities contributing to the shaping of this personalised AI system with legal, personal, social, economic and political interests, as well as worldviews and knowledge frameworks with regard to what is possible, necessary, beautiful, good, problematic, known and desired. However, not all interests are equally met in its final design.

Examining up closely the micro-cultural histories and interests invested in the individual components of a sociotechnical system, the more general patterns of sociotechnical development might appear arbitrary and at best disorganised and uncoordinated (Misa, 1988, 1992). If we look at the design of just one personal AI assistant, for example, it does not seem to be more than just that: an assistant that will help a user in a personalised manner in a specific context with a specific service. However, if we examine it in terms of the general patterns of evolution of the sociotechnical system it belongs to, we see that all components of the system, including this personal AI assistant, most often have a common direction. They belong to a map of shared systems of knowledge and meaning making in which conflicts between different interests are resolved by negotiation or, what is most often the case, domination that will then characterise their technological momentum (Hughes, 1983, 1987; Misa, 1988, 1992; Edwards, 2002).

What is important to understand here is that a technological momentum is not formed within one single societal sector or community of stakeholders by, for instance, the economic actors that invest in the technologies developed. As Hughes, for instance, describes it, while the 'mass' of a technological system is indeed influenced by the financial investment in machines, devices and structures, financial investment will not move the momentum forward alone (Hughes, 1983, p. 15). A technological momentum requires the efforts of all types of societal actors, from entrepreneurs and inventors to professional societies, organisations, businesses and governmental and educational institutions. This is also what he refers to as 'a supportive culture, or context' (Hughes, 1983, p. 140).

Returning to the example of our personal AI assistant, we can finally see it as more than just a personal assistant in digital form. In fact, one can actually discern the general cultural shape, or 'style' (Hughes, 1983) of the technological momentum that it is part of by examining the organisation of interests in its design. The personal AI assistants of the 2010s were popularised within a technological momentum characterised by leading US technology company actors as integrated components of mobiles, tablets or speakers, as with, for example, Amazon's Alexa, Apple's Siri, Google Now and Microsoft's Cortana (Bonneau et al., 2018). The 'big data technological momentum' of the products and services, such as the personal AI assistants, of these large online platforms and technology companies were throughout the 2000s and early 2010s thriving on a 'supportive culture' and 'mindset' with regard to big data in public debate, among developers and users, and even to a certain extent in policies, such as some of the EU Digital Single Market strategies, as well as individual member states' digitalisation strategies. However, as I have illustrated throughout the book, this technological momentum was in the late 2010s also in crisis, with rising critiques and revelations of, in particular, the data ethical implications of many of the big data products and services. Several critiques and concerns were raised, for example, against personal AI assistants that recognised voices and stored and processed their users' personal data in more or less intrusive manners. In particular, there were concerns regarding privacy and these services' prioritisation of the power and interests of the commercial actors behind them versus those of individuals (Chung et al., 2017; Lynsky, 9 October 2019; Maedche et al., 2019).

In the early 2020s it had become clear that the distribution of interests embedded in the products, systems and services of the large online platforms, as well as technology companies' big data technological momentum, had to transform. A 'supportive culture' of an alternative technological momentum in which the individual's interest was prioritised was emerging in public discourse; in standardisation (such the IEEE P7000s series); among entrepreneurs and inventors with different data designs and technologies (such as 'data trusts' and 'personal information management systems'); in businesses and organisations with new forms of data governance and oversight (new 'data stewardship models'); and also in legal frameworks promoting different kinds of 'data intermediaries'.

Stakeholder Interests

Now, we understand that interests are embedded in the evolution of a sociotechnical system and that these are negotiated, conflicts between them are resolved, and some interests are privileged over others when a system gains technological momentum. As a point of departure, we may consider interests in terms of different social groups or stakeholder groups. How do these different stakeholder groups contribute to the shaping of a sociotechnical system with different types of meaning creation? How do these groups actively formulate a system's successes or failures, resolve conflicts and propose solutions? Wiebe E. Bijker (1987) suggests that one of the things technology communities do is to develop and propose 'technological frames', concepts and techniques to solve problems (Bijker, 1987, p. 168). These technological frames are invested with the interests of the communities that propose them (for micro-scale scientific, personal or technological reasons or for macro-economic political and ideological reasons, among others); they shape the way in which critical problems are solved and, as a result, how the system will evolve.

A continuous appreciation of the position of the social actors that are affected by a technology is central to 'ethical governance' (Rainey & Goujon, 2011). A core mechanism here is to actively include and ensure that a multiplicity of values and views are enabled and included in the processes of technology development and deliberation. As illustrated in the context of the governance of the internet, the 'multistakeholder approach' was introduced during the World Summit on the Information Society (WSIS) process to ensure the inclusion of the interests of all stakeholders involved in the development of the internet and affected by the technology – from civil society groups and technical communities to industry and governments (Brousseau & Marzouki, 2012). However, the very act of including multiple stakeholder groups in a policy process does not, of course, ensure the fair balancing of their powers.

In the 2010s, many different stakeholder interest groups were represented in official groups and bodies and were also explicitly vocal in the public debate and politics in Europe concerning AISTIs. These groups included industry associations; consumers' national and pan-European organisations, such as The European Consumer Organisation (BEUC); digital rights NGOs, such as AlgorithmWatch, Privacy International, AccessNow, and European Digital Rights (EDRI); EU member states; national data protection agencies; EU bodies such as the European Commission's various Directorate Generals; independent EU bodies such as the European Data Protection Supervisor (EDPS) and the European Institute of Innovation and Technology (EIT); national political parties; and European Parliament political groupings. Furthermore, other less organised stakeholder groups participated, such as various user groups, independent experts, activists, journalists, academics and individual companies.

Yet the balance of these very contributions and activities of civil society groups and industry groups are often skewed by different levels of available means and resources. 'Ethical governance' requires more than an intention and act to include; it requires an understanding of the needs and limitations of different groups in society and a conscious effort to meet those needs. Now, we may continue to explore the direction and shape of a technological momentum by tracing the interests of stakeholder groups in governance initiatives, and specifically their dominance in participation and inclusion of views in official legal and policy documents and statements. However, I want to propose that interests and values do not only pertain to distinct stakeholder interest groups that can be easily discerned and categorised by examining, for example, their participation in institutional activities and settings. As we saw at the beginning of this chapter in my example of the interests influencing the work of the privacy-concerned AI developer, the complexity of their constitution is in no way easily disentangled. Interests and values invested in sociotechnical design are spread out between community 'technological frames', company goals, technological restraints, various legal quality standards, personal prejudices, needs and desires. I therefore argue here that 'ethical governance' requires a more holistic analytical view of the interests invested in sociotechnical change as a complex set of factors that come together in shared knowledge frameworks and worldviews, which can be discerned with a view to the cultures that compete and also cut across the various stakeholder groups.

Interests in Social Contexts

As a starting point, interests have contexts. They are shaped in economic and social contexts and accordingly are representative of structural power dynamics in society. As a classical sociological concept, interests are commonly considered determining factors for social action, or analytical categories for understanding societal developments (Spillman & Strand, 2013, p. 86). Interests can be discerned on a micro-level in actions of individuals and stakeholder groups, or on a macro-level in political, ideological and economic action. These 'interest-oriented-actions' are positioned towards goals and pursued as such (Spillman & Strand, 2013, p. 98). That is, they have an active agent and an object. How these agents, their goals and their object are defined, and the level of freedom of the interests in question is the result of a complexity of factors.

'Agency theory' considers how trade-offs between the different principals' interests are made by their agents, who act on their behalf in society (Spillman & Strand, 2013, p. 91). To illustrate, consider the members of the various stakeholder groups involved in the development of AISTIs; they are, as previously illustrated, represented in public debate and politics by a range of bodies and organisations. However, a member of a stakeholder group does not necessarily share the same interests as all the other members of the agent that represents them in policymaking or the public debate. Thus, an industry association's lobbying activities during the negotiation of a legal reform do not just represent 'the industry's' interest. When lobbying for the wording of a particular provision, the association represents the compromise between different members' interests.

Interests and Value-Sensitive Design

We may continue here to reflect on the distributed agency of the AI of AISTIs. Because interests in society are increasingly represented, distributed and realised in the design of partially autonomous technological nonhuman agents, it is crucial that we examine technological design as an ethically 'non-neutral' agent of interests; that is, as an agent that represents a compromise between different interests in society. Here, we can use the value-sensitive design (VSD) framework presented in the previous chapter to address the moral agency of computer design when reinforcing values of stakeholders and by design distributing the agency of these. In VSD, interests are associated with the values held by different stakeholders. These values may be reinforced or repressed by a computer technology's design.

In the 1990s, Friedman and Nissenbaum (1996) looked at different types of bias in the design of computer systems that would systematically support decision-making that unjustly benefitted or disadvantaged some groups more than others. Based on an analysis of concrete computer systems, they developed three categories to discern how bias was embedded in design:

'Pre-existing biases' come from the outside of the computer system where they 'live' in social institutions, in personal biases or attitudes held by developers. They are embedded in a computer system by explicit conscious efforts or unconsciously by institutions or individuals.

'Technical biases' emerge from 'technical constraints or technical considerations', such as limitations in hardware or software, or the use of an algorithm that, due to its context of application, does not treat all groups equally. Such biases range from imperfections in pseudorandom number generation that, for example, systematically favours those at the end of a database to, as we saw in the previous chapter, classification systems that are insufficient to represent all life in a nuanced and fully representative manner.

Finally, 'emergent biases' appear in the very context in which a computer system is used due to changes in population or cultural values, such as when a computer interface is designed for one community of users but applied in a different context, in which other users with different needs may not be sufficiently supported by the interface (Friedman & Nissenbaum, 1996, p. 333–36).

In this way, Friedman and Nissenbaum recognised that bias in computer systems is not just a technical issue but also the result of a combination of a complexity of social, technical and even individual conscious or unconscious efforts, all of which have an influence on the position and treatment of different values and interests by design. Crucially, they illustrated how a computer technology's design represents a compromise between different values held by various stakeholder groups.

Thus, an aim of VSD is to develop analytical frameworks and methodologies to resolve conflicts between the needs and values of different stakeholders in the very design of a computer technology (Umbrello & De Bellis, 2018; Umbrello, 2019). In this way, the embedded conflicts of interests of a technology are addressed as ethical dilemmas or moral problems to solve in the design and practical application of computer technologies. The approach therefore also seeks to instrumentalise the values held by different stakeholder groups, bringing them directly into the design process (Umbrello, 2019, p. 3).

By way of illustration, in the context of the interests embedded in a digital data technology, with VSD one can consider how privacy, as a value held by a stakeholder group, such as digital rights organisations, might be adversely affected by the specific design of a data-intensive technology. However, one can also suggest, as I mentioned previously, an alternative technology that is deliberately designed with privacy-preserving components (such as 'privacy-by-design'; Cavoukian, 2009) that enhance privacy values.

In the late 2010s, the idea that computer systems may have embedded bias in their design and, accordingly, potentially produce discriminatory decisions in situations where they replaced human decision-making was illustrated in a range of real-life examples of biased applications of AI. A 2020 study of patients in Boston, USA, for example, revealed how an algorithm used to score the health status of patients waiting for a kidney transplant was – by design, with the inclusion of race as a category – assigning African American people healthier scores (Simonite, 26 October 2020). Another example was the Beauty.ai beauty contest judge, which was supposed to provide the world with the ultimate measure of human beauty, but instead represented the sum of its training data by favouring light-skinned contestants. Finally, there was the facial recognition software in digital cameras that analysed pictures of people of Asian descent as people who were blinking (Mehrabi et al, 2019).

While technology design is a key focus of VSD, VSD scholars have also increasingly extended their analysis of stakeholder values to the governance contexts in which technology design is negotiated. Steven Umbrello's (2019) examination of the way in which stakeholder interests are negotiated in 'AI coordination' (the stakeholder coordination involved in what he refers to as 'beneficial AI' research and development) is illustrative of such an approach (Umbrello, 2019, p. 4). For example, he examined the multistakeholder policy process of the UK's Select Committee on Artificial Intelligence, which was appointed by the UK government in 2017 to consider the economic, ethical and social implications of AI and provide recommendations. He did so by identifying specific values (data privacy, accessibility, responsibility, account-ability, transparency, explainability, efficiency, consent, inclusivity, diversity, security and control) in the committee's evidence reports, tracing them directly to the various stakeholder groups involved in the committee (academics, non-profits, governmental bodies and industry/for profits) and ranking their order of distribution in the reports produced (Umbrello, 2019, p. 7).

Data Interests

As we have learned in the previous chapters, power is distributed in the information architectures of the Big Data Society. As follows, data can be viewed as an essential resource for the architectures of AISTIs and BDSTIs. Data becomes the locus of societal interests. This is why I think we need to be particularly vigilant of what I call 'data interests'. A 'data interest' can be defined as a motive or an intention that is transformed into data technology design that supports the agency of certain interests in the data stored, processed and analysed. I present a data interest as a motion to act on data in order to satisfy specific needs, values or goals that concern, first and foremost, data as a resource. Examples of such interests are political interests in data, com-mercial interests in data, scientific interests in data, the technical AI model's interest in data, or individuals' interest in their personal data. All of these data interests, I argue, are intertwined in the design of data technology, but also in governance activities that seek to shape the evolution of AISTIs and BDSTIs.

I propose that we examine how interests come together in general knowl-edge frameworks and values-based worldviews with enough force to shape – with standardised practices, development and adoption – the technological momentum of a sociotechnical system such as an AISTI or a BDSTI.

Let me provide two examples of data interests 'at work' in the very design and development of smart city AISTIs.

First, we have Barcelona, which was one of the first European smart city initiatives to implement a data-driven, smart city infrastructure. This consisted of an extensive Internet of Things (IoT) sensor network collecting data about, for instance, transportation, energy and air quality. It included a bicycle-sharing system with 6,000 bicycles, wireless sensors underneath roads to guide drivers to available parking spots, a waste management system with smart data-collecting trashcans, and smart lighting with sensors detecting when lights are required, in addition to initiatives such as saving energy and reducing the heat generated by old lamps (Heremobility, 2020). This is, of course, an immense network of data that also includes the data of people, sensing and acting on a mobile environment, and is something most people moving through the city would not see or feel. But it is something all big actors with an interest in data resources do see very clearly, be they commercial actors with an interest in using data to personalise, train and improve their services; scientists with an interest in improving results with data; or state actors wanting to make services and processes more efficient and control the city. There are many different interests in the data resources of the smart city AISTIs of Barcelona. The main risk here is that only a few interests of the most powerful are met in the very data design of the city. However, in 2015, the new mayor, Ada Colau, took the smart city initiative in a new direction, together with the city's Chief Digital Technology and Innovation Officer Francesca Bria. Their mission was to develop the city's data infrastructures 'for and by the people'. They developed a digital transformation agenda for Barcelona that views 'data as commons', opening up data to help the city's entrepreneurial ecosystem, including SMEs in the ICT sector, and empowering citizens with tools that allow them to selectively disclose the information they would like to share (Heremobility, 2020). Barcelona, together with Amsterdam, also became a pilot city in the European DECODE project, which developed smart city initiatives and tools where citizens could choose how and with whom they share their data.[1]

The second example of the role of data interests by design that I want to use here is the centralised CityBrain AI data system developed by the Chinese tech giant Alibaba. It monitors every vehicle in the city of Hangzhou, China, and has helped reduce traffic jams greatly; however, it also does numerous other things. The system constantly monitors video footage of traffic, looking out for signs of collisions or accidents to alert the police. It combines data from the transportation bureau, public transportation systems, a mapping app and hundreds of thousands of cameras. In this way, not only are accidents automatically detected and responded to faster, but also things such as illegal parking are tracked live (Beall, 30 May 2018). Of course, there are also interests in the centralised data system of CityBrain, just as there are in the data system of Barcelona. However, the key difference between the two smart cities is that the CityBrain AISTIs do not have citizen oversight or control baked into their design. The huge amount of data generated by Hangzhou's system is designed to meet the interests of, first and foremost, a few power actors, namely law enforcement, the Chinese government and the private company Alibaba.

In the 2010s, various technological cultures with different priorities in terms of meeting certain patterns of interests in data by design – as with these two smart cities examples from different parts of the world – were competing on a global arena for the momentum that frames the practices that go into designing the AISTIs and BDSTIs of the era. Here I propose that we need to be critically aware of these interests in the very design of AI, to understand what kind of power we enforce when we design the components of data intensive sociotechnical infrastructures: democratic powers, monopolistic powers, authoritarian or totalitarian powers. Because this is what we do. We create, provide and distribute power by design.

To help forward an exploration of data interests in AI design and development, I have elsewhere (Hasselbalch, 2021) presented five clusters of themes based on key data metaphors: 'data as resource', 'data as power', 'data as regulator', 'data as vision' and 'data as risk'. I described these as follows (here adapted and revised from the original publication):

Data as resource
Who or what provides the data resource? Who or what has an interest in the data resource? How is the data resource distributed and how does the human being benefit?

The 'data as resource' cluster concerns the different interests in the distribution of data resources of the data design. If data is a resource, it can also be separated from that which it represents (a person or an artefact). Data can be 'provided', 'accessed', 'gathered', 'labelled', 'extracted', 'used', 'processed', 'collected', 'acquired' and 'put' into a system in a 'structured' or 'unstructured' way. A resource is a corporeal and spatially delineated thing, something we can be in possession of, place in containers or create boundaries around, store and process, and it is something we can be with or without. The resource metaphor is common in public discourse on big data (Puschmann & Burgess, 2014).

Technology critic Sara M. Watson (n.d.) refers to the dominant metaphors for personal data in public discourse as 'industrial', as if it were a 'natural resource' to be handled by 'large-scale industrial processes'. Mayer-Schönberger and Cukier (2013) depict data as the raw material of a big data evolution of the industrial age. But the 'data as resource' metaphor actually denotes two different things. One type of resource is indeed the raw material of industrial processes that is processed and turned into products. Another type of resource is the kind that makes us stronger as individual human beings. Being 'resourceful' also means that you as an individual have the capacity, and the physical, psychological and social means. The first type of resource mentioned here is tangible and material, the other is social and psychological. However, both are what the linguistics scholar George Lakoff and Mark Johnson in their classic book *Metaphors We Live By* (1980, p. 25) would refer to as 'container' metaphors: entities with boundaries that we can handle and reason about.

Subsequently, data as a resource can be protected and governed in very tangible ways, and in each case micro- and macro-stakeholder interests are involved in these governance and resource handling frameworks. Generally speaking, industries have an interest in the raw material of their data-based business models: political players have an interest in providing rich data infrastructures for AI innovation to compete in a global market; the engineer has an interest in volumes of data to train and improve the AI system; individuals have an interest in protecting their personal data resourcefulness or even enhancing their data resources by creating their own data repositories and benefitting directly from these (as the 'data trust' or 'personal data store movement' advocates). Obviously, treating the psychological and social resources of individuals as material resources in an industrial production line represents a conflict of data interests; but in addition, informational asymmetries also create very tangible social and economic gaps between the data rich and the data poor, which is a conflict of interest on a more general structural level in society. In this way data is not only a 'raw material' of AI technologies, data is also, and essentially, the individual's resource.

Treating data as the resource of the individual means preventing harm to individuals through the protection of data. Personal data protection means protecting the social and psychological resource of human beings, or said in other words, protecting 'human dignity as well as mental and physical integrity' (HLEG A, p. 12). To provide an example from the introduction of this chapter, an AI developer will make use of different data resources for the design of an AI system. These data resources have to be either processed on her own computer or she can use the more powerful cloud-based AI platforms of, for example, Google or Amazon. However, by doing this she also has to share data resources with these actors. If prioritising individuals' data interests, when dealing with personal data she has to trust that these companies do, in fact, also prioritise the treatment of data as a personal resource of the individual human being. Trusting these actors thus entails a data interest design choice. Other actors may also have interests in the data resource. For example, if the developer is creating an AI system for assessing the performance of employees, the manager that she designs the system for might have an interest in creating a bigger data resource that can assess minute details of the employees' workday. The manager might even want to collect data from outside the work place, for example, from the employees' social media presence, to enable the AI system to predict potential risks to the workplace, such as internet searches for jobs outside the company or organisation. The employees, on the other hand, have an interest in keeping the data resource to a specific limit, and a union representing these employees might also want to get involved to safeguard the data interests of the people they represent.

Data as power

Who is empowered or disempowered by data access and processing? Does the data design support the human responsibility, power and data agency? How are conflicts between different data interests resolved to the benefit of the human being?

The second cluster, 'data as power', is closely connected with 'data as resource', as the distribution of resources also constitutes a distribution of power. A society is based on balances of power between social groups, states, companies and citizens. A democratic society, for example, represents one type of power structure in which the governing powers are always balanced against the individual citizen's power. We can here think of power dynamics in terms of 'informational power' and information asymmetries between individuals and the institutions and companies that collect and process data in digital networks. Access to or possession of data can be associated with the dominance of certain societal groups, institutions and/or businesses, but also with the very function of democratic and/or 'fair' processes, where access to information and an explanation of data processing is the basis of 'accountability' and 'fairness'.

Basically, what we need is data design that supports 'human agency' via informed decisions and choices'. For example, access to the data and data processes of a system can provide an investigative journalist or researcher with the power to challenge an AI system. The journalist or researcher has an interest in the data of the system that may be inhibited by a company's interest in keeping its algorithms and data design proprietary. The researchers involved in the Machine Bias study mentioned earlier, for example, did not have access to the calculations used for defendants' risk scores. At one point they received the basics of the future-crime formula from the company behind, Northpoint, but the company never shared the specific calculations, which it said are proprietary (Angwin et al., 2016).

Data as regulator

Which interests does the implementation of law in the data design serve and prioritise? Does the data design of the AI system enhance the values of the law? How are interests between different legal frameworks resolved by design?

Technology design is a type of 'regulator' that either protects or inhibits law (Reidenberg, 1997; Lessig, 2006). The 'data as regulator' cluster emphasises the role of data design in the legal implementation and realisation of law in society. In an ideal situation, law and technology design supplement each other enhancing the values of the law, though in reality data design often just complies with legal requirements, and at times the very properties of a data technology may even be in direct contrast with law. The data intensity of AI can, for example, directly challenge legal principles such as data minimisation, privacy and data protection by design, or the right to information and/or an explanation. Different stakeholder interests in the data design may also be in conflict with legal frameworks, such as the fundamental rights framework, in which the data interests of individuals are primary values. For example, non-democratic state actors have an interest in data for social control purposes; certain types of business have interests in tracking and collecting data to enhance their data-based business models without consideration of individuals' rights, or similarly, scientists might have interests in improving their research with big data analytics without these considerations.

Data as vision

Can human beings 'see' the data processes and their implications? What does the data design see (the training data) and then perceive (how is it instructed to act on the training data)?

Vision is the very agency of data interests in the data design (what we are able to see or not, and how we see it). In an ideal constellation, vision is an effortless extension of our eyes. However, in a digital data-based environment, the instruments (our digital eyes) that we use to see and perceive our environment are quite literally extended into a data design that constitutes a management of what we can actually see.

As previously discussed, data design also functions as a moral agent, in that it prescribes and manages our active engagement with the information it handles, as well as the digital information infrastructure in which it exists. Eyes and vision are metaphorically embodied in data systems. While data is often described as AI's sensory system[2] on which it develops its mode of action, as the EU High-Level Expert Group on AI's ethics guidelines on trustworthy AI states, data is also the 'eyes' of the individual human being. Data will, for example, 'yield the AI system's decisions, including those of data gathering and data labelling as well as the algorithms used' (HLEG A, p. 18). The concept of 'transparency' is here often used in the most literal sense to denote our ability to see the data processes that should be documented and made traceable. 'Black box' algorithms' (Pasquale, 2015) processing of data may in this context blur our vision or make us blind to the reasoning of an AI system. The management of visibility, that is, the very architecture of visibility of emerging technological environments, can be said to constitute a mode of social organisation and distribution of powers (Brighenti, 2010; Flyverbom, 2019). What is made visible, what remains invisible and, importantly, who is empowered to see through the social organisation of visibilities directly influences the agency of interests in society. The eyes, as the data design of an AI technology, do exactly that. To provide an example, we might create an AI system to analyse the data of people on social welfare benefits. This may have a dashboard for the public institution's social workers, which provides general data statistics, fraud detection and risk scores on individuals. The dashboard is our eyes within the AI system. In this case, only the social worker's data interest has eyes and so does the public institution's interest in controlling public resources and optimising work processes. But we could also think of other types of data design in which the people on social welfare have eyes through data access, and hence agency to, for example, add and correct flawed data or personalise the services provided to them.

Data as risk

To whom or to what could the data design be a risk? Who or what has an interest in preventing and managing the identified risks? How are conflicts and/or alignments between identified risks to the human being's interest in managing the risks (including its responsibilities towards its environment) resolved?

Risks are a contemporary concern in politics, business conduct and public discourse in general. The sociologist Ulrich Beck described a preoccupation with risk prevention and management in his seminal book *The Risk Society* (1993) as an uncertainty produced in the industrial society, and as the result of a modernisation process in which unpredictable outcomes are emerging and accumulating (Beck, 1993). 'Risks are not "real", they are "becoming real"', Beck later proclaimed when describing a development of concern with 'world risks' (Beck, 2014, p. 81). They take the shape of 'the anticipation of catastrophe' and their management as an 'anticipation of further attacks, inflation, new markets, wars or the restriction of civil liberties'. Importantly, the depiction of a risk 'presupposes *human decisions*, human made futures (probability, technology, modernisation)' (Beck, 2014, p. 81).

The data of AI is laden with potential risks that we want to prevent and manage. As described in the four other metaphorical clusters, data is generally associated with the management of risks to society's resources, to democracy, to the rule of law and to agency through visibility. But in this last metaphorical cluster, data is a risk per se that must be anticipated and managed. Criminals can for example 'attack' the data of an AI system and 'poison' it, data can 'leak', it can be 'corrupted by malicious intention or by exposure to unexpected situations' (HLEG A, p. 16), and data's carbon footprint poses a risk to the environment when concentrated and processed in data centres with high CO_2 emissions. If we continue with the notion that risks are not 'real' per se but based on our own predictions about possible future scenarios, then we may also assume that their proposed management and prevention is the product of interests and motives. On the AI development side, an AI engineer training an AI technology will have an interest in the risks posed to the quality and accuracy of the training data, while a data protection officer will consider risks posed to identified individuals with a data protection impact assessment. On the AI deployment and adoption side, an individual will also have a 'data-as-risk' interest in keeping their personal data safe from unauthorised access, while an anti-terror intelligence officer's risk scenario involves the detection of terrorist activities and might therefore consider the end-to-end encryption of a data design a risky design choice. The consideration of different 'data-as-risk' scenarios guided by the various interests involved in these will direct disparate design choices that might be aligned but might also be in conflict.

Cultural Data Interests

We can think about data interests in terms of the most traditional stakeholder groups that have interests in the data generated, processed and stored in AISTIs and BDSTIs. Multistakeholder internet governance initiatives, for example, will seek to ensure the inclusion of stakeholders generally from four respective communities; government, the private sector, civil society and technical communities. I want to examine interests in a different way, that is, to try to understand how data interests take form as cultural interests that cut across these different stakeholder groups. I do this to illustrate how power is not just a uniform interest expressed in coherent group formations. In fact, power is expressed in a complex system of meaning making, 'styles', 'world views', or what we in this chapter will explore as 'cultures'.

Take, for example, the power dynamics of the policy debate on AI and data in the late 2010s that was often presented as a clear-cut conflict of interests between civil society and industry stakeholders. For example, the industry interests manifested in the EU High-Level Expert Group on AI's ethics work were described by one group member, in an op-ed in a German newspaper, in terms of the 'group's extreme industrial weight', which, according to this member, had directly influenced the creation of 'a lukewarm, short-sighted and deliberately vague' set of ethics guidelines (Metzinger, 2019). Nevertheless, being a member of the group myself, and with many years of experience of participating in multistakeholder initiatives like this, I perceived some different power dynamics. Although I did indeed see the traditional conflicts of interests between civil society group members and those of certain industry members in some discussions and in the details of the group's work, as I will describe later, I also saw the emergence of a cultural interest in a 'human-centric' approach to AI in response to the critical social moment that I have explored in previous chapters. This cut across the different stakeholder groups and was spelled out as the 'European third way': a unique point of reference of the group's ethics guidelines' and the AI policies later proposed by the European Commission, with an emphasis on the European Fundamental Rights legal framework as a point of departure. I will return to a more detailed analysis of this later. My point is here that depictions that only see the traditional 'stakeholder group'

interests rarely grasp the nuances and complexities of the cultures and interests invested in governance processes. That is, as argued before, even within
specific stakeholder groups there are power struggles between different technological cultures. A company driven by a big data mindset could, for instance,
be in direct conflict with another company motivated by ideas on privacy by
design, and whose interests may therefore align better with the interests of
a digital rights civil society stakeholder group. This is why I argue that we need
an investigation of more general cultural patterns of power negotiation and
positioning that very often form alliances across the more traditional interest
formations and stakeholder groups, such as 'the industry', 'civil society' and
'state'. This is also why I argue that traditional multi-stakeholder governance
approaches may be inspired by an 'ethical governance' approach to consider
the explication and negotiation of values as a component of stakeholder inclusion in policymaking.

Let us here take 'data ethics' as an example of a shared cultural interest
that developed across different stakeholder groups in the 2010s in Europe.
Philosophy and media studies scholar Charles Ess (2014) has illustrated
how culture plays a central role in shaping our ethical thinking about digital
technologies. For instance, he argues that people in Western societies place
ethical emphasis on 'the individual as the primary agent of ethical reflection
and action, especially as reinforced by Western notions of individual rights'
(p. 196). Such cultural positioning in a global landscape could also be identified in the European 'data ethics' policy debate, in which the European 'inner
demons' (Bauman, 2000) of an epoch and the corresponding ethical response
were voiced time and again. Take, for example, the way in which one member
of the European Parliament, in a debate in Brussels on the data protection legal
reform, describes the issues at stake:

> It is all about human dignity and privacy. It is all about the conception of personality
> which is really embedded in our culture, the European culture ... It came from the
> general declaration of human rights. But there is a very, very tragic history behind
> war, fascism, communism and totalitarian societies and that is a lesson we have
> learned in order to understand why privacy is important.[3]

'Human dignity', the first article of the European legal framework on fundamental rights, is an ethical principle with historical and cultural roots in the
experiences of totalitarian regimes of power and the cruel and undignified
treatment of one community of human beings in Europe during the Second
World War. In the 2010s, what I have referred to as 'data ethics spaces of
negotiation' were in Europe also cultural endeavours to protect and urgently
transfer European cultural values, such as human dignity and privacy, into
technological developments (Hasselbalch, 2020). As stated in an EDPS report

from 2015: 'The EU in particular now has a 'critical window' before mass adoption of these technologies to build the values into digital structures which will define our society' (EDPS, 2015, p. 13). European 'data ethics public policy initiatives' (Hasselbalch, 2019) took form as spaces of negotiation over particular 'cultural values'. Policy- and decision-makers positioned themselves against what was perceived as a pervasive, opaque threat to a distinctly European set of values. This was a threat that was understood to be embedded in sociotechnical design and business conduct that worked as a 'wrecking ball', aiming not simply, as one President of the European Parliament said in a speech in 2016, to 'play with the way society is organised but instead to demolish the existing order and build something new in its place'.[4] The values embedded in technology design and business conduct were here in particular associated with a new form of power. As the then Director for Fundamental Rights and Union Citizenship of the European Commission DG Justice Paul Newitz claimed in a 2017 public debate: 'The challenge is with the controllers, [who] have power, they have power over people, they have power over data, and what are their ethics? What are the ethics they instil in their staff? In house compliance ethics? Ethics of engineers?' (Nemitz, 2017).

The European 'third way on AI' that I mentioned before took form in the late 2010s as an emphasis on shared cultural values cutting across various European stakeholder interest groups. AI had, on a global scale throughout the decade, increasingly been embedded in the private and public sector infrastructures of emergency, health care, finance, security, defence, law, e-government, transportation and energy. The US was a first mover in terms of global capital investment in the development of an AI ecosystem, and China rapidly followed suit (Merz, 2019; Lapenta, 2021). So by the late 2010s, EU decision-makers were realising that AI had become an area of strategic importance, transforming critical infrastructures in all the aforementioned sectors, and was therefore also a driver of economic development. I have in the article 'Culture by Design' (Hasselbalch, 2020), described how the contours of an institutionally framed 'EU AI agenda' took shape as a distinctive cultural positioning on a global market with an emphasis on 'ethical technologies' and 'Trustworthy AI', as follows (here adapted and revised from the original publication):

The EU AI agenda's 'cultural positioning' was spelled out in policy documents and statements in a process that involved EU members states, the EU High-Level Expert Group on AI, a multistakeholder forum called the European AI Alliance and, predominantly, various different Directorates of the European Commission. At the same time, the EU increased its annual investment in AI development and research and established an agreement to join forces with national strategies on AI in member states. Described as a way to ensure Europe's competitiveness on a global scale, the EU's AI agenda was also often presented in this period as a response to a 'global AI race' in the public media, debates and reports. The main focus here was the competition among regional players for global leadership on the resources for AI (for example, data access), capital investment, AI technical innovation and practical and commercially viable research and education, as well as 'ethics' as a form of risk mitigation and regulation (Merz, 2019). However, besides a race for resources, technological supremacy and risk mitigation, the explication of a values-based cultural framework for AI also played a key role in defining a shared interest in AI among various stakeholder groups (Hasselbalch, 2020).

The European Commission published its first communication on artificial intelligence in early 2018, accompanied by a declaration of cooperation on artificial intelligence signed by 25 European member states (which was detailed later in 2018 in a Coordinated plan on artificial intelligence, 'Made in Europe'). This preliminary approach to AI focused on cooperation among member states, multistakeholder initiatives, investment, research and technology development. Predominantly, AI was described here as part of a European economic strategy within a global competitive field. The values-based positioning was not a core strategic element of this first communication, but only hinted at: 'The EU can lead the way in developing and using AI for good and for all, building on its values and its strengths' (European Commission K, 2018) and a first step to addressing ethical concerns was made with the plan to draft a set of AI ethics guidelines.

Then, the EU High-Level Expert Group on AI (HLEG) was established, with 52 selected members consisting of individual experts and representatives from different stakeholder groups with the core task to develop AI ethics guidelines, as well as policy and investment recommendations for the EU. From the beginning, the group's work was defined within a distinctive European framework. As evidenced by a European Commission representative's comment at the group's first meeting in Brussels: 'AI cannot be imposed on us', and concluding that 'Europe must shape its own response to AI' (HLEG E, p. 4)

The 'European response' was already here defined in terms of what was described as a shared set of European values. For example, at the same meeting, the chair introduced the core constituents of the group's mandate and the European Commission's expectations of the group as follows: 'It is essential that Europe shapes AI to its own purpose and values, and creates a competitive environment for investment in AI' (HLEG E, p. 2). This proclamation was later included in the discussions of the group and defined as the search for a distinctive European position in a global setting: 'Discussion also centred on identifying the uniqueness of a European approach to AI, embedding European values, while at the same time identifying the need to operate successfully in a global context' (HLEG E, p. 5).

'European values' were also the basis of the ethics guidelines published a year later, in April 2019. Here, values were introduced with reference to the European Commission's vision to ensure 'an appropriate ethical and legal framework to strengthen European values' (HLEG A, p. 4). The key references were the European rights-based legal frameworks, such as the Charter of Fundamental Rights and the General Data Protection Regulation. However, European values were also covered by the unifying ethics framework, defined as a 'human-centric approach' with an emphasis on 'human dignity' and in which the individual human being's interests prevail over other societal interests:

> The common foundation that unites these rights can be understood as rooted in respect for human dignity—thereby reflecting what we describe as a 'human-centric approach' in which the human being enjoys a unique and inalienable moral status of primacy in the civil, political, economic and social fields. (HLEG A, p. 9)

As described before, several ethics guidelines for AI were published in European member states, outside Europe and by international organisations. Most notably, only a few months after the HLEG's ethics guidelines were published in 2019, 42 countries adopted an OECD recommendation that included ethical principles for 'Trustworthy AI'. In comparison with these other more principle-based ethics guidelines, the HLEG's ethics guidelines were particularly focused on the operationalisation of ethics in the very design of AI and on providing concrete and practical guidance to AI practitioners. Thus, in the end, it was the delineation of a specific type of technology design and culture of AI practitioners that became the ethics guidelines' unique cultural positioning. The European 'third way' to AI in this way came into being as an 'ethics' of the design of AI, or what the HLEG named 'Trustworthy AI'. This also meant that when working with the policy and investment recommendations that were published in June 2019, the high-level group proposed Trustworthy AI as a core European strategic area (HLEG B, 2019).

Simultaneously, the European Commission's brief 'concern' with a European approach to AI in an early strategy transformed into a strategic point of positioning. The first coordinator of the HLEG, Nathalie Smuha, has described how the work of the HLEG was quickly adopted within the European Commission's general AI strategy (Smuha, 2019). At that time, there were 700 active high-level expert groups tasked with drafting opinions or reports advising the Commission on particular subjects. However, their input was not binding, and the Commission was independent in the way it took into account the work of these groups (Smuha, 2019, p. 104). Nevertheless, when the HLEG on AI presented the ethics guidelines to the Commission in March 2019, an almost immediate agreement was reached to publish the last communication in the two-year period – 'Building trust in human-centric AI' (European Commission N, 2019), which stated its support for the seven key ethics requirements of the guidelines and encouraged all stakeholders to implement them when developing, deploying or using AI systems.

This culminated in a promise by the new president of the European Commission, Ursula von der Leyen, at the end of 2019: 'In my first 100 days in office, I will put forward legislation for a coordinated European approach on the human and ethical implications of Artificial Intelligence' (von der Leyen, 2019). And in early 2021 the world's first regulatory proposal targeted at AI was published, with an emphasis on 'European values', the fundamental rights and safety risks specific to AI systems, and the key requirements of the HLEG's ethics guidelines, reflecting 'a widespread and common approach, as evidenced by a plethora of ethical codes and principles developed by many private and public organisations in Europe and beyond, that AI development and use should be guided by certain essential value-oriented principles' (European Commission O, 2021, p. 8).

The European Cultural Interest in the Data of AI

We can think about technological change as constituted by power dynamics, as a negotiation between interests, and as a question of power and very human interests in dominion. According to Hughes (1983, 1987) a dominant culture is the very shape of a technological momentum and this means that competing cultures and their invested interests must convert to the most powerful culture of the momentum or perish (Hughes, 1983).

We have looked at how a European AI Agenda evolved in the 2010s into a distinctive European cultural positioning in a global environment with an emphasis on 'ethical technologies' and 'Trustworthy AI'. We can also think of this AI agenda as a political interest in shaping a global technological AI momentum. Let us here take a closer look at this cultural interest as a 'data interest' of what was framed as a European 'data culture'. In the article 'Culture by Design' (Hasselbalch, 2020), I presented four cultural components of this (here adapted and revised from the original publication):

The first cultural component of the European AI agenda was the delineation of a European cultural context for a technological momentum in the 2010s and early 2020s. The HLEG's policy and investment recommendations (HLEG B, 2019) for example described the different phases of digitalisation in European societies where AI forms a 'third wave':

Europe is entering the third wave of digitalisation, but the adoption of AI technologies is still in its infancy. The first wave involved primarily connection and networking technology adoption, while the second wave was driven by the age of big data. The third wave is characterised by the adoption of AI which, on average, could boost growth in European economic activity by close to 20 percent by 2030. In turn, this will create a foundation for a higher quality of life, new employment opportunities, better services, as well as new and more sustainable business models and opportunities. (HLEG B, 2019, p. 6–7)

The European cultural context was delineated in various ways. First and foremost, with a comprehensive regulatory data protection reform, the EU had established itself as what was often referred to on the global arena as a 'regulatory superpower' in the digital field. The GDPR and the EU Charter on Fundamental Rights' legal frameworks for AI innovation were thus identified as Europe's distinctive risk-based approach. In addition, the European stakeholders that constitute a 'technological momentum' became a central topic of the European policy negotiations and debates on AI. This included, for example, a focus on AI practitioners, scientists, entrepreneurs, data analysts, educators, the workforce, policymakers and citizens in general. Centrally, a European data infrastructure for AI also gradually took form. Data was often described as the main driver for the technological AI momentum. In fact, the third 'AI' wave of European digitalisation, it was argued, would be 'driven by the age of big data' (as described by the HLEG). Very early on the European Commission recognised, in its first communication on AI, that data was a key factor for the development of AI in Europe with a reference to the creation of 'data-rich environments' as 'AI needs vast amounts of data to be developed' (European Commission K, 2018). The EU was consequently also described as 'a pivotal player in the data economy' (HLEG B, 2019, p. 16) as data 'is an indispensable raw material for developing AI' (HLEG B, 2019, p. 28). Therefore, data was also held to be core to what the stakeholder interests of the AI momentum were invested in: 'Ensuring that individuals and societies, industry, the public sector as well as research and academia in Europe can benefit from this strategic resource is critical, as the overwhelming majority of recent advances in AI stem from deep learning on big data' (ibid.).

The second cultural component of the European AI agenda consisted of a negotiation and delineation of the very foundational cultural values and ethics for 'European AI'. This was predominantly again also described with reference to existing European legal frameworks, such as the General Data Protection Regulation and the Charter of Fundamental Rights. Nevertheless, it was also recognised that, with the rise of data intensive technologies such as AI, a process of cultural meaning negotiation had been initiated exposing and seeking to resolve conflicts of interests that existing legal frameworks did not seem to be able to solve. As described previously, the European AI agenda therefore also explicated a 'human-centric approach' as a foundational values-based framework, stressing that the human interest prevails over other interests. It also involved a particular approach to data governance that empha-sised the empowerment of individuals in the handling of their personal data. For example, the HLEG's ethics guidelines outlined a clear framework for the management of data with one of the seven requirements, 'privacy and data governance', specifically addressing the human-centric values embedded in the data design of an AI technology. Here, the concept of human agency was correlated with the individual's knowledge and the information provided for the individual to make decisions and challenge automatic systems (HLEG B, 2019, p. 16). The human-centric approach was first and foremost developed as an overarching framework for resolving conflicts between different interests and values embedded in AI innovation, for example, between the interests of an ethics-driven and a data-driven approach to innovation, or the clashing interests between machine automation of labour and the human work force. A range of 'human-centric' solutions were thus suggested to resolve these conflicts, for example, the development of mechanisms for the protection of personal data and individuals to control and be empowered by their data (resolving conflicts between citizen data empowerment and state/business data power); ethical technology as a competitive advantage (resolving conflicts between ethics and data-driven innovation); humans-in-the-loop/command AI solutions for the workplace and upscaling the AI skills of the workforce (resolving conflicts between automation and the replacement of workers); and generally focusing on the use of non-personal data in business-to-business (B2B) AI solutions rather than the personal data of business-to-consumer (B2C) solutions (resolving conflicts between risks of using personal data and the data intensity of AI technology development).

The third cultural component of the European AI agenda consisted of the explication of a European 'technological culture' challenging in particular the 'politics' and 'values' as well as power asymmetries embedded in existing AI technology and design. Thus, the skills, the education, the methods and prac-tices of what was referred to as 'ethical technology' were core to the European AI agenda discussions. This was also when a European 'ethical design' culture grew into being as *the* European position of the global AI momentum. The European strategic investment in a particular 'ethical' 'technology culture' of AI was also an essential focus of the HLEG's policy and investment recom-mendations. It was stated that Europe needed to 'foster understanding' and 'creativity' and generally 'empower humans by increasing knowledge and awareness of AI' (HLEG B, 2019, p. 9–10)

The fourth cultural component of the European AI agenda involved the depiction of a cultural data space, a European data infrastructure for AI. At the beginning of the 21st century, 'data cultures' were created on the basis of an interjurisdictional digital flow of data. The very 'architecture' of a global data infrastructure had emerged as an interjurisdictional space challenging

European data protection/privacy values and legal frameworks. For example, at a very early stage, the European Court of Human Rights (ECHR) had to consider, on several occasions, the challenges that the progress of digital technologies posed to the ECHR's territorial definition of jurisdiction in cases concerning the right to privacy and the legal uncertainties this caused.[5] This also meant that AI was initially developed on the basis of an interjurisdictional and territorial global big data infrastructure. The revelations of embedded data asymmetries in the form of surveillance scandals, fake news and voter manipulation had provoked a European concern with foreign 'data cultures' and their data architectures. The European AI agenda therefore proposed an alternative European data-sharing infrastructure for AI based on a foundational values-based approach to data, but which was also confined within the European jurisdiction and geographical space. In the policy and investment recommendations, the HLEG described data infrastructures as the 'basic building blocks of a society supported by AI technologies'. These data infrastructures were also described as the foundation of a European AI critical public infrastructure, and therefore should be treated as such: 'Consider European data-sharing infrastructures as public utility infrastructures.' The development of this European space should also be invested with a specific set of values and designed 'with due consideration for privacy, inclusion and accessibility, by design' (HLEG B, 2019, p. 28). This is also where the cultural interest in data stood out. The values-based approach was conceived as a cultural effort to transfer European values into technological development, positioned against a 'non-European' threat pervasively embedded in technological infrastructures: 'Digital dependency on non-European providers and the lack of a well-performing cloud infrastructure respecting European norms and values may bear risks regarding macroeconomic, economic and security policy considerations, putting datasets and IP at risk, stifling innovation and commercial development of hardware and computer infrastructure for connected devices (IoT) in Europe' (HLEG B, 2019, p. 3).

2.   CULTURE AND TECHNOLOGY

Thomas P. Hughes (1983) considered cultural environments a crucial component of the fourth phase of a sociotechnical system's evolution in which it gains momentum. In fact, a technological momentum is created by the prevalence of a dominant culture. It is this very common force that brings together all the diverse factors of human, social and technical character to create a technological momentum: 'Taken together, the organisations involved in the system can be spoken of as a system's culture' (Hughes, 1983, p. 15). Hughes identified 'cultures of technology' as 'contextual elements' for the growth of a technological system that arise from the inside of the system in the shape of values and ideas of engineers and systems builders, but also from the outside in the form of cultures of a regional power (Hughes, 1983, p. 363).

Thus far, I have used the term 'culture' in the way that Hughes uses it (and it is used in Science and Technology Studies [STS] in general) to describe shared conceptual and material frameworks for the development of sociotechnical systems. Moreover, I have used the term 'values' to designate the ethical and moral dimension of developments (as is done in value-sensitive design [VSD]). However, what does it actually mean when we say that policymakers or engineers, data practitioners and systems builders share and practice a cultural values-based framework? What does it mean to have a shared culture that is forceful enough to create a technological momentum and make a system grow and consolidate in society?

Technological Style and Cultural Values

With a VSD approach to technological development, the moral evaluation of what we consider 'good' and 'ideal' becomes part of the technical design process. Values are in fact 'idealized qualities or conditions in the world that people find good' (Brey, 2010, p. 46). However, these values are not just the personal ideals of individuals working with the design of a technology; they are also intentionally advanced by various types of stakeholders with shared interests and distinct shared cultures. Culture is the foundation of ethical evaluation, and culture is shared.

Hughes describes different regions as distinct settings for various ways of conceptualising and designing technology. Culture is here represented in the 'technological styles' of the regions. He illustrates how differences in 'technological styles' became particularly apparent in the 20th century due to the increasing availability of 'international pools of technology' (including, for example, international trade, patent circulation, the migration of experts, technology transfer agreements, and other forms of knowledge exchange; Hughes, 1987, p. 69). As he says, technological style is an 'adaption to environment' (Hughes, 1987, p. 68), the technological language, so to speak, of the culture of the economic and social institutions involved, in which knowledge and practice are systemised and conceptualised. In this way, even ethical evaluation in technological development can be argued to be a product of the culture and interests involved.

Sociologist Epstein (2008) has illustrated how the study of the concept of culture in science and technology settings evolved in two ways; from examining culture inside of the institutions and scientific labs to studying its role in the outside world of adaption and consolidation. Early studies of cultures inside scientific institutions brought forward the notion of knowledge as a cultural product, and thus also brought with them a focus on the competition of scientific actors in pursuit of interests within distinct cultural environments (Epstein, 2008, p. 168). Scientific credibility and authority were therefore also considered 'cultural resources' with an emphasis on the very negotiation processes in which claims are made within scientific institutions (Epstein, 2008, p. 168). Thus, the very systems and networks of meaning making that frame technological practice became a key focus of cultural analysis. In this context, culture was later examined as a multiplicity with an emphasis on distinct scientific cultures with, for example, ethnographic studies in science labs (Epstein, 2008, p. 169). Sociological accounts of science and technology, Epstein argues, on the other hand identify culture outside of institutions and scientific labs in conceptions of material culture and politics created by humans to organise human life, such as with the technological means and modes of modern state governance. In the outside world, cultural consequences are also examined in practices of boundary creation or classifications of the world, which take form as modes of social ordering and power distribution (Epstein, 2008, p. 172–3).

More generally in STS, culture has been related to the way we get to know things and the skills and resources we use to create a technology.[6] Thus, in the book *Science as Practice and Culture*, science and technology scholar Andrew Pickering (1992), for instance, defines culture in a footnote as a resource for doing scientific work:

> Throughout this essay, 'culture' denotes the field of resources that scientists draw upon in their work, and 'practice' refers to the acts of making (and unmaking) that they perform in that field. 'Practice' thus has a temporal aspect that 'culture' lacks, and the two terms should not be understood as synonyms for one another: a hammer, nails, and some planks of wood are not the same as the act of building a dog kennel – though a completed dog kennel might well function as a resource for future practice (training a dog, say). (Pickering, 1992, p. 3)

Distinct 'knowledge cultures' or 'technological cultures' can also be described as environments with rules for a technology's design and adoption in society. The sociologist Harry M. Collins, one of the key people behind the sociology of scientific science studies at the British Bath School, defines 'cultural skills' as intents and purposes and sets of implicit socialised rules of action for the design of a technology (Collins, 1987, p. 344). He considers scientific skills as belonging to different 'explicable' or 'inexplicable categories'. There are the formal facts and rules, the 'heuristics' ('rules of thumbs') and the manual perceptual skills that are visible and may be explained, and then there are the cultural skills that he describes as the inexplicable or 'hidden' components of technology development (Collins, 1987, p. 337). These are required in order to use and understand formal facts and rules, heuristics and manual skills to develop technology. However, they are silently shared within communities and only acquired by the ones within the same cultural community. Thus, to an outsider without the cultural skills set, a crucial framework is missing, which is why in cases where different communities come together, the cultural component must be explicated. He explains this as follows:

> When we interact with those who are close to us in cultural terms, we make do with few explicit remarks, and these convey meaning because so much is shared at the outset. But, as cultural and contextual distance increases between communicators, the potential ambiguity of more and more messages becomes realized, and more needs to be made explicit and can be made explicit. (Collins, 1987, p. 344)

Consequently, Collins looks at the way in which skills within the different categories may transform and transfer into other categories, such as when a rule of thumb becomes a fact and formal rule written down in a manual. Crucially, he holds that the transformation of skills from one category to another involves changes in 'cultural ambience' which is 'enmeshed in wider social and political affairs' (Collins, 1987, p. 344).

Collins actually also uses AI expert systems to illustrate how implicit cultural skills may be transformed into explicit skills. With an AI expert system, all the skills of a human expert – the explicated formal facts and rules, as well as the inexplicable cultural skills – will have to be coded into the system for it to act 'intelligently'. This may be done more easily, he argues, with human experts such as solicitors and medical specialists, who have skills that already rest on stores of codified information, rather than with specialists who rest on less organised expertise and cultural skills (Collins, 1987, p. 344).

This theory of cultural skills and skills transformation is an excellent example of the role of culture as a crucial component of technological development and change; that is, as the invisible and often taken-for-granted resources, skills and conceptual frameworks that are nevertheless crucial for technological development and change.

An AI system will only be as intelligent (and useful in its cultural setting) as its cultural design. Hence, failure to incorporate the invisible cultural component into an AISTI's design will also imply malfunction in its consolidation in society. In a complementary manner, we may here additionally use a VSD approach to consider culture as a component of the very act of moral evaluation within the process of designing values into a technology, so to speak. Also, failure here to incorporate culturally sensitive moral evaluation into the very design of AISTIs will mean a clash with the ethical and moral evaluation of a given culture and society. Hence, as we can see, STS and VSD approaches place an equally strong emphasis on the role of culture and shared cultural frameworks for sociotechnical development. However, neither of these perspectives offers a conceptualisation of culture as such.

What is culture? How do we identify and discern the particularities of the cultural component and the cultural ethical evaluation component of technological change? In fact, how can we constructively understand the very qualities and characteristics of the different cultural systems for making sense of the world, which are held up against each other in the power struggles of sociotechnical change as more than just conflicts and battles between systems? How do we constructively make sense of their compromises? I here turn to cultural studies to examine the concepts of culture and cultural values in more detail. I do this to offer a way to make the cultural shape and structure of a technological momentum visible in order to guide it reflectively with a data ethical governance approach.

Culture and Power

At this point we understand that culture is a type of value system that brings together communities with shared conceptual frameworks and resources. Culture is also an active system in the sense that it constitutes specific priorities, goals and ways of organising the world that are actively imposed in society when practiced by, for example, engineers and represented in material things (such as our technological systems). However, crucially, we can also conclude, based on the previous discussion, that culture is constructed and invested with interests, and thus, the dominant cultures of specific communities and societies are only one view of the world. On those grounds, culture is like Bowker and Star's classification systems (2000), or vice versa, the classification system is a cultural system of representation (I return to this point in more detail), practiced as if it was complete, but in effect a reduction of the world into one set of categories that do not represent the complete picture. Something is always left out. Something does not belong. That is, culture as a cultural system is never complete.

In 1958, one of the key founders of the British Cultural Studies tradition, the Marxist theorist Raymond Williams, famously defined culture as a 'shape', a set of 'purposes' and 'meanings' that are expressed 'in institutions, and in arts and learning' and in the 'ordinary' (Williams, 1958/1993, p. 6). Culture, he argued, is more than just the refined, curated and selected art and literary works produced by one social class; it is also 'popular', expressed in mundane everyday practice. It is 'a whole way of life' (ibid.). Culture consists of prescribed dominant meanings, but critically also their negotiations. Crucially, Williams argues that the meaning of culture is in 'active debate and amendment under the pressures of experience, contact and discovery' (ibid.) and as such it is simultaneously 'traditional' and 'creative'. Hence, there are two sides to culture: 'the known meanings and directions, which its members are trained to' and 'the new observations and meanings, which are offered and tested' (ibid.).

In this perspective, we may also consider culture a site of power negotiation between the 'state of affairs' and the potential for change.

Williams' account of culture as 'ordinary' and a 'whole way of life' is essential to the British cultural studies tradition that emerged in the 1960s and 1970s from the Birmingham School, with a specific focus on the study of popular culture and subcultures (Agger, 1992). Traditional elitist and exclusive conceptions of culture were replaced with studies of the culture of everyday life of the working class (Thompson, 1963/1979) and youth subcultures, introducing issues of race (Gilroy, 1987/2012; Hall, 1990/1994) and gender (McRobbie, 2000) and their representation and construction. That is, culture is not just one; it is multifaceted, it is institutionalised but also mundane, subcultural – informally and formally created by and in interaction with people, including minority groups, and artefacts – and the meanings of these cultural relations are never stable. They are, from the outset, constructed systems of meaning making and therefore always up for contestation and social power negotiation.

Is there a 'science' for these cultural systems of sense-making? How can we make sense of the role of culture as more than just a black box that silently and invisibly defines practices and the design of our sociotechnical infrastructures? In semiotics and semiology, cultural practices, products and representations are investigated as components of systems of cultural meaning making that are interrelated and gain meaning through their systematic ordering. Rules of linguistics are applied to culture to formulate a 'science of signs', but it is not only the organisation of words in systems of signs and meaning that represent who we are and how we experience the world in culturally constructed systems. All types of sign systems are examined as language and, accordingly, as cultural representation.

One of the world's most famous semioticians Roland Barthes, for instance, found cultural meaning in the mythologies of the French bourgeoisie expressed in everything from wrestling matches to advertisements for soaps (Barthes, 1957/1972). Myths, he argued, are 'systems of signification' in which we make meaning of things in the world as elements of cultural discourse and the cultural order of specific moments in history: 'for a myth is a type of speech chosen by history: it cannot possibly evolve from the "nature" of things' (Barthes, 1982/2000, p. 94). Everything, he argued, is culturally meaningful, including technological artefacts that have cultural systems of meaning making embodied in their very design. By way of example, he proposes that French toys of the 1950s were designed to generate a microcosm of adult life and the roles and functions society encourages adults to perform: the post office, the school, the army, the medical profession, and dolls that urinate and need to be taken care of by little girls. As such, they were cultural signs of the bourgeoise adult culture actively reproducing its patterns of social ordering (Barthes, 1957/1972, p. 53–5).

Along these lines, culture theorist and political activist Stuart Hall defined culture as 'systems of signification', 'conceptual maps' and 'maps of meaning' (Hall, 1997). These are the conceptual systems in which we organise, classify, arrange and relate concepts with each other (Hall, 1997, p. 17). The cultural maps of meaning ensure that we understand each other and act coherently in a community when interpreting the world in the same manner. Cultures are, according to Hall (1997), social systems in which meaning is actively created and shared. Culture is not imposed on us from above. We actively practice, learn and belong to a culture. In this way, culture is also an active habit of sharing codes to communicate and make sense of the world, and these codes can be traced in cultural products that are interrelated in 'systems of representation'. Specifically, shared codes are the keys to cultural systems of signification as they are the 'stabilisers' of meaning (Hall, 1997, p. 21). Based on 'unwritten agreement', 'conventions of representation' and 'cultural "know how"', the codes of culture are also the key to cultural belonging, enabling us to 'function as culturally competent subjects' (Hall, 1997, p. 22).

In his essay 'Encoding/Decoding', Hall (1980) describes processes of cultural meaning production as an interaction between moments of coding and decoding cultural messaging. A television, for example, which was the key popular communications technology of the 1980s, he argues, is encoded with a cultural 'dominant cultural order' with 'preferred readings' for their decoding receivers (Hall, 1980, p. 123–4). Although the process of meaning making is not 'symmetrical', according to Hall, and might even be distorted or misunderstood, the very moment of meaning making is embedded with the preferred readings that constitute a core component of the 'conditions of perception' (ibid., p. 119–121). Whether reflective and intentional or unintentional and habitual, these practices of encoding and decoding cultural systems are constitutive of culture and, importantly, reproduce the dominant cultural order (Hall, 1980, p. 123). Hall writes:

> These codes are the means by which power and ideology are made to signify in particular discourses. They refer signs to the 'maps of meaning' into which any culture is classified; and those 'maps of social reality' have the whole range of social meanings, practices, and usages, power and interest 'written in' to them. (Hall, 1980, p. 123)

As we may recall, a range of VSD and STS scholars have also considered technologies 'non-neutral' cultural products embedded with 'values' and 'politics'. Like Barthes' toys and Hall's television, Langdon Winner, for instance, views technologies as linked with the power dynamics of a given society and asks:

> Does this state of affairs derive from an unavoidable social response to intractable properties in the things themselves, or is it instead a pattern imposed independently

by a governing body, ruling class, or some other social or cultural institution to further its own purposes? (Winner, 1980, p. 131)

In their book *Data Feminism* (2020), the data scientists and feminists Catherine D'Ignazio and Laura F. Klein illustrate how, through time, predominantly male data science cultures have been sustained in work environments where female data scientists' work was unappreciated and unrewarded. These data science cultures, which the authors describe as *oppressive*, are not only a gender struggle; they set the goals and priorities of the very data design in which power is distributed and where often minority groups' interests are repressed; for example, when minority groups are either underrepresented in data used as the basis for decisions made on social benefits, or when critical scientific medical analysis only benefits one privileged group, or on the other hand when a minority group is overrepresented in data in such a way that puts them at a disadvantage in society, such as data from specific city zones used for predictive policing.

In addition, in the very data science teams that develop the data design of the digital information architecture of our daily lives, D'Ignazio and Klein see an underrepresentation of minority groups. For instance, according to an AI Now report, women comprise only 15% of AI research staff at Facebook and 10% at Google (referenced in D'Ignazio & Klein, 2020, p. 27). These oppressive data science cultures are reflected in real-life experiences of minority groups working in data science and reflected in the data technology and design that have become an increasingly ubiquitous component of our social environment (D'Ignazio & Klein, 2020). Just as I also propose, D'Ignazio and Klein therefore use the concept of 'power' as the axis for the injustices they see reflected in data science practices and data design:

> We use the term power to describe the current configuration of structural privilege and structural oppression, in which some groups experience unearned advantages – because various systems have been designed by people like them and work for people them – and other groups experience systematic disadvantages – because those same systems were not designed by them or with people like them in mind. (D'Ignazio & Klein, 2020, p. 24)

A number of critical data studies are directly addressing the power dynamics of the environments and technological cultures that frame the practices and design of specifically AI and big data technological developments like these (O'Neil, 2016; Eubanks, 2018; Noble, 2018; D'Ignazio & Klein, 2020). However, critiques of the distribution of power in technological cultures of science and technology practice go further back. Since the late 1970s, a distinct research field counting feminist technoscience scholars, such as Judith Butler, Donna Harraway and Sandra Harding, raised critiques of science and technology practices and knowledge in terms of the cultural gender power dynamics they reproduce and enforce (Åsberg & Lykke, 2010). Technology and science are here considered sites of dominion and identity struggle in which repressive categories of gender are fortified. Science is produced within scientific knowledge environments characterised by traditional gender roles, creating opportunities for some while repressing others. These are environments of repression that are then reinforced in the very technology created. The result is that existing power relations and dynamics in society are reinforced and, in many cases, even amplified.

Understanding technology as a cultural product, and examining how technological practice is embedded in often repressive socially ordering cultural systems of meaning making that are lived and experienced by individuals, compels us to consider the cultural component of sociotechnical development as a specific object of our ethical scrutiny. This is also why the cultural systems of technology and technological practice per se are relevant as ethical problems that we should seek to solve with an applied ethics approach.

Following this, a core concern we may have regarding the data ethics of power as applied to AISTIs involves their constitution as cultural systems of a type of social ordering in which the interests of dominant actors in society have the primary advantage while other minority interests are further disadvantaged. This is where we might also consider technology as a highly specific potential site of rebellion and freedom. Donna Harraway's critique of the construction and positioning of gender through science and technology practice was, for example, voiced in her momentous 1985 Cyborg Manifesto, in which she imagined an alternative information architecture, a union between human and machine, based on 'socialist and feminist principles of design' that replace 'the informatics of domination' (Harraway, 1985/2016, p. 28).

Data Cultures

I wish to further elaborate the conceptualisation of data cultures, the technological cultures that frame data science and practice, with a view on culture and power. Data cultures are the culturally coded conceptual maps of the engineers, data scientists, designers, deployers, legislators and users of data systems. As I have argued throughout this book, these are not always shared, even within specific stakeholder groups and communities, and they may even be in conflict. Furthermore, they certainly are, as we saw previously, interrelated with societal power negotiation and struggle.

Computers are in essence information in a particular form. At face value, a computer is an information system that technically enables different types of data collection, sharing and processing for purposes that are supposed to be useful to humans. Information is processed, managed, modelled and classified according to mathematical formulas created by computer information scientists. The applied science of a computer scientist, or what I refer to here as a 'data designer', consists of creating smoothly and efficiently working computer information infrastructures that collect, store and organise data and, if modelled as an AI system, sense, are trained and evolve on data to make 'intelligent' decisions. However, as I have illustrated in previous chapters, AISTIs and BDSTIs are, of course, not just that. They have politics, values and a delegated moral agency that has 'non-neutral' social and ethical implications or, said in other words, they have distinct 'data cultures'.

As a first step, we can define a data culture as one that is constituted by cultural practices in the form of ethical evaluation and choice when developing a data design (e.g., coding, labelling, managing, collecting and selecting data). Data scientists and data designers' practices are framed within cultural systems of meaning making and as such they are active practices of (reflective or non-reflective) ethical evaluation. Or put differently, the very practice of developing a data system and design is a cultural practice (Acker & Clement, 2019). Geoffrey C. Bowker, for instance, examines the cultural constitution of biodiversity databases, illustrating the non-neutrality of 'raw data'. The data cultures of the data scientists' practices that go into these databases are, he says, also practices of the 'layering of values' into a data infrastructure. In this way, the very database in which data is stored and accessed is, from the outset, a site of 'political and ethical as well as technical work' (Bowker, 2000, p. 647). Thus, not only are data scientists' practices cultural, data is culture. It is not a natural given; it is not a raw material that may exist in itself and by itself (Bowker, 2014). Data has no meaning outside the cultural system, the database or the data practice.

This 'non-neutrality' of data and data design is not in any way particular to an AI or big data information system. As Geoffrey C. Bowker and Susan Leigh Star (2000) have illustrated in detail, the information classifications and standards that are essential to any working infrastructure have, throughout human history, actively organised human relations with social and ethical implications for the people involved – from the classification of tuberculosis patients for purposes of incarceration in asylums to race classification during apartheid for purposes of segregation (Bowker & Star, 2000). Thus, there is always an 'ethical dimension' to practices of collecting, organising and processing data:

> We have a moral and ethical agenda in our querying of these systems. Each standard and each category valorises some point of view and silences another. This is not inherently a bad thing – indeed it is inescapable. But it is an ethical choice, and as such it is dangerous – not bad, but dangerous. (Bowker & Star, 2000, p. 5–6)

Information classifications and standards are two different types of cultural practice within a data culture. One is the very practice of segmenting the world according to certain criteria, while the other is the institutionalisation of this practice of segmentation.

To start with, a classification is described by Bowker and Star as a 'set of boxes (metaphorical or literal) into which things are put to then do some work – bureaucratic or knowledge production' (Bowker & Star, 2000, p. 10). Classifications are consistent, have unique classificatory principles, and are mutually exclusive, which also means that when sorting information, you can only adhere to one classification system or another. There is, so to speak, no 'outside' of the classification system: 'The system is complete' (ibid., p. 11). Of course, a perfect classification system is not possible in practice, as there will always be ambiguity or disagreement as to whether an object belongs in a specific category. Furthermore, the information used to place an object within a specific category is in reality never complete. Human life, for example, does not fit easily into one classification system, and when it is reduced to one set of categories it will not represent a complete picture. In the previous chapter, I used Alpaydin's example of human age, which is not easily placed in a 'box' by an expert system as we are not just 'old' or 'young', but are all ageing gradually (Alpaydin, 2016, p. 51). In the context of AISTIs, we also saw in the last chapter how one of the key drivers for the evolution of AI – from the 1970s expert systems to the increasingly autonomous big data machine learning systems of the 2010s – was to overcome the limits of expert systems in representing their environments by building systems that could pro- gressively perceive their nuances. It is exactly the ideal of the completeness of the computer algorithm's classificatory work that creates the ethical and social implications when applied to human life. The computer algorithm is designed as a complete classification system and does not understand itself beyond this ideal of completeness. Yet, it is never complete and if deployed as such on human life it may, exactly for this reason, have grave ethical and social con- sequences (which was also one of Cathy O'Neil's [2016] primary concerns).

For example, a predictive computer algorithm that makes risk assessments on the potential future criminal acts of an individual (such as the COMPAS algorithm), may process various types of personal data regarding an individual to make these risk assessments. For instance, this could be data about the indi- vidual's location, correlated with data on crime rates in different areas of a city. The individual might be living in an area with high levels of crime, which the computer algorithm then classifies as high risk. Because the computer algo- rithm's principles for classification are always complete in theory, no other data about the individual is used to nuance the placement of this individual in a 'high-risk' category. If the algorithm is then deployed in a judicial system as a 'complete system', this may have grave social consequences for individuals living in areas with high crime rates.

As a core practice of an information scientist's work, classifications are embedded in the infrastructure of their working environments, invisible and shared in smaller communities. However, they may also become standardised and institutionalised, shared in more than just one community. Bowker and Star describe standards as: 'any set of agreed upon-rules for the production of (textual or material) objects' (Bowker & Star, 2000, p. 13). Standards are created to make things work smoothly together and they are enforced by, for example, legal bodies, the state or professional organisations. Although no 'natural law' mandates their existence (and they therefore come into existence as the result of social negotiations among various stakeholders), they are heavily institutionalised and thus difficult to change (Bowker & Star, 2000, p. 14). Standards are, for example, key components of a well-functioning infrastructure and vice versa (Dunn, 2009). However, it is not only technical components that need to function in a standardised manner in order to work efficiently together with other components of a standard; the very work practices of people also need shared cultural systems of meaning making to function well. As Bowker states elsewhere: 'Working infrastructures standardise both people and machines' (Bowker, 2005, p. 112).

In the context of AISTIs, one example of institutionalised technical standards that shape the work of an AISTI data designer is the International Organization for Standardization's (ISO) systems requirements standards. ISO is an international standard-setting body with representatives from national standards organisations. It develops technical safety and quality standards for the development of products and systems, which it certifies to ensure they meet the requirements of the standards. ISO standards are internationally recognised, shared and developed. In this way, they ensure consistency in the design of a product or system, that the safety requirements are met, and that they are compatible with other products and systems that are compliant with the same standard.[7] The list of current technical standards for IT is long and diverse, spanning standards on IT security to information coding.[8] Although ISO standard certification is not a legal requirement, a certification of a system or a product does help ensure legal compliance.

In Europe, the General Data Protection Legal Reform (GDPR) presented a legal framework that was integrated into standards for the design of information technologies that process personally identifiable information. For example, the ISO/IEC 27701, published in 2019, specified requirements 'for establishing, implementing, maintaining and continually improving a privacy-specific information security management system' with mapping to the GDPR. While the core purpose of the GDPR – that 'The processing of personal data should be designed to serve mankind' (Regulation [EU] 2016/679,

p. 2) – was not very different from the 1995 data protection directive that it replaced, the GDPR did have a stronger emphasis on the quality of the actual technology design processes and practices set in place to ensure data protection and the rights of the individual.

To illustrate this with an example, the new provision on 'data protection by design and default' (Regulation [EU] 2016/679, article 25) required that privacy and data protection were considered and specifically designed into IT systems from the start and not as an afterthought. This meant, for example, that what we have previously referred to as the 'big data mindset' (Mayer-Schönberger & Cukier, 2013) was challenged, as one key task for a data designer would be to consider data minimisation as a core quality of the design of a data technology. Other technical design requirements included the pseudonymisation of personal data, the creation of IT design with greater built-in transparency in regard to the functions and processing of personal data, and the ability for individuals to monitor the data processing.

Standards are centrally controlled and maintained, but of course they are not unequivocally embedded in practice. They are understood, interpreted and transformed by practitioners; negotiated in institutional, social and cultural contexts; and deviations are even accepted (Bowker & Star, 2000, p. 13–15). For example, the GDPR does not provide technical specifications. It is not a technical standard; it presents the legal compliance framework for technical and organisational measures and has a twofold aim to protect the fundamental rights and freedoms of individuals while ensuring the free movement of personal data within the EU (Regulation [EU] 2016/679). Data designers may in fact technically implement this legal objective with a weight on either one or the other side of this aim. A data designer with a 'big data mindset' will, conceivably, seek to fulfil the legal provisions on data protection only for reasons of legal compliance, but not as a quality standard for the very design of the technology. However, for another type of data designer, data minimisation and individual privacy could be a quality design goal in and of itself.

In the 2010s, a growing design and business movement was developing information technologies with privacy and data protection as quality criteria for its work. In our book *Data Ethics – The New Competitive Advantage* (2016), Pernille Tranberg and I described a range of examples of data designers and companies that presented their interpretations of the quality criteria for personal data processing with declarations of independence, manifestos and public statements describing privacy and data minimisation as a quality standard for their work (Hasselbalch & Tranberg, 2016, p. 91). For example, the data designer Aral Balkan told us: 'It is possible to build systems where individuals have ownership and control of their own data, on their devices, instead of holding it in a cloud where a corporation has ownership and control' (Balkan in Hasselbalch & Tranberg, 2016, p. 92).

Another CEO of a German toy company, Vai Kai, whose main product was a set of Internet-connected wooden dolls, Matas Petrikas, considered his customers' privacy the basis of all design and innovation decisions. Therefore, Vai Kai's internet-connected dolls did not have a camera or microphone like most other such dolls on the market at that time. He told us that privacy by design was a quality standard for his IT product:

> We think about privacy as a value all the time. It is part of our conversation. I assume other companies would never have had the conversation we had during our development phase that led to the conscious decision not to include a microphone. (Hasselbalch & Tranberg, 2016, p. 98)

We can think of these emerging practices and ideas about data innovation as cultural paradigm shifts, when one cultural 'normality' of the knowledge and practice foundation of information and computer science practices transforms into another. Here, we want to look at the moments when what is taken for granted about designing a computer information system is uprooted, when old meanings are challenged, and something different and new takes their place. Formal technical standards are difficult to change, but they are not mandated by laws of nature. They represent a dominant culture's values-based quality criteria. Thus, while the cultural ethical evaluation of one data culture is formalised in standards, it is not set in stone. The kind of ethical evaluation that takes place when one data culture meets another with different values and quality criteria (for example, on handling data interests) is the most critical one. Standards have, time and again, been changed and updated in contexts of critical social negotiation. These changes to the standards for data design practices will represent, as well as require, paradigm shifts and new normalities. New priorities are set, new guidelines for practices are created, and new 'scientific imaginations' and 'worldviews' – to use the expressions of Kuhn (1970) – emerge.

This is why we can also see the regulatory reform of the data protection legal framework in Europe as a symptom of a paradigm shift in the cultural environment of the information computer scientist/data designer – just as the 'data ethical movement' among designers and companies that Tranberg and I described (2016) represented this shift. ISO standards in information technological practice were increasingly also updated to reflect a new normality for the data designers' work. For example, a range of new standards for the development of AI were developed and published by the ISO/IEC JTC 1/SC 42 Committee on Artificial intelligence. Several of these were, when I examined them in early 2020, concerned with the big data design of AI, and many were also specifically focusing on the social implications of AI, such as ISO/IEC AWI TR 24027 on 'Bias in AI systems and AI aided decision making';

ISO/IEC PRF TR 24028 on an 'Overview of trustworthiness in artificial intelligence'; and SO/IEC AWI TR 24368 on an 'Overview of ethical and societal concerns'. Another standard-setting organisation that has created standards reflecting the new, more 'ethically' reflective cultural environment of the AISTI data designer, was the IEEE Global Ethically Aligned Design for Autonomous Systems P7000 series of standards, with standards such as P7002 on the 'Data Privacy Process of AI', P7006 on 'Personal Data AI Agents', or the P7012 standard for 'Machine Readable Personal Privacy Terms', among many others.

All in all, changes of the data cultures of BDSTI and AISTI development in the 2010s involved a paradigm shift in what was considered the normality of data and information knowledge and science practices, that is, changes in the 'normal' practices and 'normal' standards and requirements. This is how it works. Real change of direction of a sociotechnical infrastructure means a fundamental change in the cultural ways of working with information, how it is collected, processed, stored, analysed and used, which also means a change in the scientific imagination that shapes these processes. One step in the change in the data cultures has been based on the realisation of the incompleteness of a system's data design when applied to human life.

Critical Cultural Moments

Data cultures may be stable and institutionalised, but as we have seen, they are also negotiated and challenged. I have previously described the spaces of negotiation that happen on a macro-scale of time, in a moment of crisis, just before a large sociotechnical system's consolidation. However, moments of critical ethical evaluation also happen in data design practices and when data designers are negotiating and making design choices in a micro-design context, as I have illustrated before, for instance, with reference to the negotiation of 'data interests' in AI design. Information scholar Katie Shilton (2015), for example, studied the values expressed by an internet architecture engineering team – the Named Data Networking project – and created a taxonomy of types of values of the data designers, associating them with different interests: '(1) those that respond to technical pressures and opportunities; (2) those focused on personal liberties; and (3) those influenced by an interest in the collective concerns of an information commons' (Shilton, 2015, p. 8).

The moments when alternative cultural systems of meaning making are proposed and confront existing cultural normalities is what I call 'critical cultural moments'. These moments are of crucial importance for an applied data ethics of power approach. First of all, they are moments in which cultural meaning making within a data culture is destabilised. We therefore want to preserve them to enable human critical engagement with sociotechnical change. However, we also need to understand that the most peculiar thing about these moments is that they are ephemeral, they do not last. They tend to dissolve in the dominant data cultures of data design, laws and standards. A data design is, for example, not just 'coded' data; it is data culture in code. It performs a specific data culture. That is, data design has cultural properties that can be examined as culturally encoded systems of signification. In fact, the system's data design is *culture in action*, enacting a data culture's values. Said in other words, a dominant culture is reproduced without challenge when transformed into seemingly neutral technical 'professional code' (Hall, 1980, p. 126). As outlined by Collins (1987) in his description of cultural skills and AI, in expert systems culture is transformed into explicated categories, literally coded, and in advanced self-learning systems even encoded within the systems when autonomous machine predictions and decisions are made. Here, AI systems like these are a specific challenge to a data ethics of power, as they will never be able to 'understand' like a human (Searle, 1980, 1997); or in other words, the system is not capable of reproducing the *critical* cultural moment of meaning production. The cultural classification of the world is actively coded and produced within the system. This means that the cultural moment of meaning production between an encoded 'preferred meaning' and the decoded negotiations of meaning is moved into the system. It also means that the critical cultural moment, when it conflicts with alternative data cultures that would otherwise have emerged had the interpretation been made in a qualitative, spatially and temporally situated context and by humans, never takes place.

This is a tricky problem, because the critical cultural moments of meaning making I propose are the most *human* components of sociotechnical developments. They have human characteristics, in the sense that they occur when human 'memory' and 'intuition' (I will return to these concepts in the last chapter) are sparked. They must therefore also be preserved in the data cultures that shape the development of AISTIs and BDSTIs. To do this we need to prioritise in very practical ways the human interest in the data of big data and AI infrastructures via the meaningful involvement of human actors in their very data design, use, governance and implementation.

NOTES

1. The DECODE project: https://decodeproject.eu
2. The EU High-Level Expert Group on AI for example described AI's sensory system as: 'perceiving their environment through data acquisition, interpreting the collected structured or unstructured data, reasoning on the knowledge, or processing the information, derived from this data and deciding the best action(s) to take to achieve the given goal' (HLEG A, p. 36)

3. Marju Lauristin (26 January 2017). MEP debate: The regulation is here! What now? [video file] https://www.youtube.com/watch?v=28EtlacwsdE
4. Martin Schultz (3 March 2016) Technological totalitarianism, politics and democracy https://www.youtube.com/watch?v=We5DylG4szM
5. See an analysis with key case law references at https://mediamocracy.files.wordpress.com/2010/05/privacy-and-jurisdiction-in-the-network-society.pdf
6. I do recognise that culture is also a contested concept in STS, for example, as represented in the debate between Callon & Latour, 1992 and Collins & Yearley, 1992.
7. Described on the ISO website: https://www.iso.org
8. Described on the ISO website '35 Information Technology': https://www.iso.org/ics/35/x/

5. What is data ethics?

'Mankind lies groaning, half crushed beneath the weight of its own progress. Men do not sufficiently realize that their future is in their own hands.'
Henri Bergson, 1932

When the data protection legal reform was implemented in Europe in 2018, the risks and implications of the big data era were already household items in the public news cycle. The Cambridge Analytica scandal was one such thorny news item effecting waves of uproar in European media, policy and beyond. It revealed a British consultancy firm's use of social media and big data analytics methods, based on the machine learning analysis of the data of 87 million people worldwide, including 2.7 million Europeans, to influence democratic processes in the US and UK (Stupp, 6 April 2018). Expositions such as these – data leaks and hacks, algorithmic discrimination and data-based voter manipulation – kept public attention tuned in to the data ethics implications of everyday online life, politics and culture in the late 2010s.

In public policymaking, discussions regarding data ethics had therefore equally gained traction and, as I have illustrated throughout this book, numerous public policy initiatives under headlines such as 'data ethics', 'Trustworthy AI' and 'ethical technology' were created in member states and in European intergovernmental policy contexts. In particular data ethical implications were equated with the new structures of power of the sociotechnical big data systems.

As such, European policy and decision-makers were increasingly positioning themselves against a pervasive, opaque form of power embedded in BDSTIs and AISTIs dominated by 'GAFA', an acronym for the four leading US big data technology companies: Google, Apple, Facebook and Amazon. Thus, in late 2019 and early 2020, the first steps were made to implement a European 'Trustworthy AI' and 'ethical technology' policy agenda in the shape of European laws and cultural values with, for instance, the publication of a strategy for taking back control over the European data space and regaining data sovereignty and even a potential ban on facial recognition AI (Delcker & Smith-Meyer, 16 January 2020).

Then, one morning, we went into lockdown. A pandemic was sweeping across the planet, and the EU and governments all over Europe, all over the world, were scrambling to control, mitigate and predict the evolution of the crisis with various means and modes of governance. This included the swift introduction and adoption of several data-based digital technology and/or AI-based solutions. In fact, Europe in general experienced an immense digitalisation and AI boost, such as smart working and education platforms, telemedicine, contact-tracing apps, big data-based algorithms to support diagnosis and epidemiological studies, personalised medicine, and care robots (Craglia et al., 2020).

When contact-tracing apps were being developed in countries all over Europe, the debate on privacy and the choice between a centralised and decentralised management of data became fierce. However, it did not go further than discussions about privacy. In actuality, the BDSTI power of Google and Apple was only cemented even further when the companies blocked the development of several contact-tracing apps in European member states (Hasselbalch & Tranberg, 20 May 2020). At the same time, Europe saw an acceleration of technologies that were handling personal data while prioritising safety and public health, including drone surveillance, location tracking, biometric bracelets, facial recognition and crowd behaviour analysis (Craglia et al., 2020). One may even argue that, in the face of the COVID-19 crisis, only some of the 'data ethics' concerns regarding the distribution of power in the age of big data that had matured in Europe over the previous decade remained, while others were set aside (Martens, 2020; Vesnic-Alujevic & Pignatelli, 2020).

What happened to the power of data ethics in 2020 when COVID-19 swept through Europe? How could some core concerns be swept aside with it and only some remain? In this last part of the book we look at human power and data ethics and present a formative framework for a data ethics of power. My proposition is here that a data ethics of power cannot be put aside, nor can it only be applied when considered useful. That is, we can formulate data ethics guidelines, principles and strategies, and we can even program artificial agents to act according to their rules. However, to truly ensure a human-centric distribution of power, data ethics has to become more than just a moral obligation, a set of programmed rules, it has to be 'human'. This means that data ethics has to take the form of culture, to become a cultural process, lived and practiced as a way of being in the world. As such, a data ethics of power first and foremost addresses the cultural conditions and structures of power, rather than the sole value properties of technology design.

There are a few premises for a data ethics of power that I must outline first. Some are based on the conclusions of the previous two parts of the book that dealt respectively with the power of BDSTIs and AISTIs. In essence I here considered BDSTIs and AISTIs as representing particular types of ethical problems and accordingly proposed data ethics as a response to these specific challenges. I argued that human agency and experience exist in the contemporary structures of BDSTIs and AISTI power. I described how a complex of cultural powers are immobilised in BDSTIs and enacted in AISTIs. As such, moral agency is increasingly also a property of AISTI agency and external to human agency. This does not mean that humans are not still involved. We design technology, we use it, we interpret it, we shape it in our own image and according to our interests, but the very agency of our ethical evaluation and agency in particular is increasingly externalised.

This constellation is a critical concern of a data ethics of power, because only humans can have the type of critical ethical agency that a data ethics of power requires. A data ethics of power is a revolt against a 'closed' exclusive society. It seeks an inclusive 'open society' based on love without a specific interest (Bergson, 1932/1977; Lefebvre, 2013) and the multiplicity of culture, and culture as a whole way of life (Williams, 1958/1993). Therefore, what I have referred to as 'critical cultural moments' and 'spaces of negotiation' are required to challenge the immobilised dominant culture of data systems.

A data ethics of power addresses BDSTIs and AISTIs' power structures for human agency and experience as core problems. Yet, a delineation of problems does not answer the key question of this final part of the book: What is data ethics? To answer this question, I propose an answer to two quintessential sub-questions: Why is a data ethics of power important? How can a data ethics of power achieve the 'good society'? The answers to these two questions constitute the formative framework for a data ethics of power, illustrated in Figure 1. I will briefly outline the answers here and then go into detail in the rest of the chapter.

1. Why do we need data ethics?

First, I want to understand why we need a data ethics of power. What ontology necessitates a data ethics of power? What are the premises of our being in the world, and accordingly, from an ethics perspective, what constitutes a 'good society' and 'being'? These questions lead to the first formative component of a data ethics of power, which is an ontology that I will describe with reference to the philosopher Henri Bergson's 'process ontology' and the development of this by the philosopher Gilles Deleuze.

Ontology: Data ethics is a way of being in the world. It is an ontology of process and movement, where life is only stable and fixed when represented in systems of meaning making (Bergson, 1903/1999). Agents act in the world with different capacities. Humans are one type of agent, while technological agents, such as AI, are another. They both have agency, but not the same, as there are fundamental differences (Searle, 1980, 1997; Smith, 2019; Amoore, 2020; Pasquale, 2020). Therefore, in sociotechnical environments there are also two different ethical potentials: an AI agent can indeed be said to have a rational intellect and also act with moral agency (as illustrated in Chapter 4), but it does not have the human ability to 'think movement' (Bergson, 1907/2001, p. 318); it does not have the 'semantics' (Searle, 1980, 1997), 'doubt' (Amoore, 2020), 'judgment' (Smith, 2019) or even 'expertise' (Pasquale, 2020), and it can therefore never be an ethical agent by itself. Even to act as a moral agent of human ethical agency, the AI agent needs human empowerment, which is created by ensuring the 'critical cultural moments' in design and adoption. This human empowerment is in essence what I consider the human approach of a data ethics of power.

2.   How can a data ethics of power achieve a 'good society' (an open society)?

We need an understanding of the ethical problems of our age. These are the BDSTIs and AISTIs in which power, culture and moral agency are captured and stabilised and in which the essence of our being to evolve and recreate ourselves in a constant process (and accordingly an open society) is immobilised. The core ethical problems of the Big Data Society are a precondition for the second formative component, the action-oriented approach, which will create the conditions for the critical human ethical agency that is necessary to achieve an open society.

Practice: Data ethics is a form of critical applied ethics that explores the conditions of power in the sociotechnical systems of the Big Data Society to actively create and ensure ('data ethical governance') 'spaces of negotiation' and 'critical cultural moments'. *Spaces of negotiation* are spaces carved out in society with a material presence in which values and interests are exposed and negotiated. Their core objectives are critique and negotiation. They are possible when 'systems' (material/immaterial and technological/ cultural) clash and controversy arises. For example, in policy, spaces of negotiation are inclusive initiatives established to negotiate values and establish shared ethical frameworks. However, they are only viable in moments where specific conditions make critical value negotiation possible (Hughes, 1983, 1987; Moor, 1985). *Critical cultural moments* have special human characteristics. They emerge and are only possible when human memory and intuition are privileged and provided the time and space to tinker. For example, in AI design and adoption, the critical cultural moments are constituted by the level and type of human involvement and prioritisation of human environments in the technical design and adoption of AI systems.

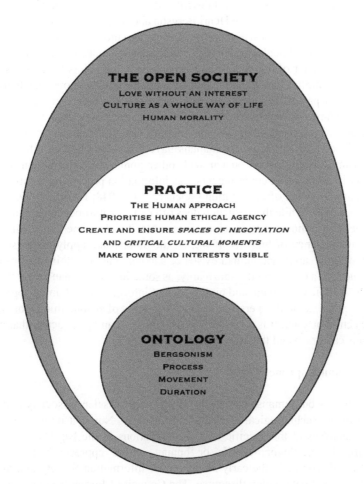

Figure 5.1   A formative framework for a data ethics of power

1.   WHY DO WE NEED DATA ETHICS?

When asking this question, I am in effect, with my answer, also making a statement about the ethical capacities of human and nonhuman agents in sociotechnical data infrastructures, as well about their respective status and relationship. What do we assume about human beings, their environments, technologies and crafts? How is the relation between technology and humans, human agency and technological agency perceived within a data ethics of power? The answer lies in what I describe as a human approach, which I base on some foundational ideas of Henri Bergson about humans and our capacities and place in the world. I will therefore also go into more detail on some of his key critiques and thoughts on society, morality and humans in this first section of the last chapter of the book. Summarising the main points here, Bergson's perspective implies that while the human approach does acknowledge humans as a natural part of the physical world, it rejects the idea that humans and their nonhuman agents have the same ethical capacities, and as follows, the same ethical responsibilities. We can think of this in the context of information science and ethics, where humans have been described as information 'organisms' and 'objects' on par with other nonhuman information agents that act in information environments – as information processing components of a material world of information (Wiener, 2013/1948; Floridi, 1999, 2013; Bynum, 2010). While this depiction does offer a less anthropocentric and more holistic view of humans and their environments, it at the same time challenges the ethical agency of humans. If we, on the other hand, apply a Bergsonian view, we may approach human existence and being as something that cannot be reduced to a represented stable reality, as something other than the reception of information, processing and giving back of information. A human approach in data ethics can in this perspective also be thought of as something more than mere action on data. It is a way of life, a glimpse of an open society that only humans can grasp and therefore be responsible for.

The Human Approach

An emphasis on humans (humanism) in theories on science, society and the world of our technological artefacts is no novelty. Nor is it unique in more recent analyses of the specific ethical implications of the Big Data Society. Specifically, the 'human-centric' or 'human-centred' approach had a revival in policy discourse in the early 2000s on the Information Society and in the 2010s' AI and data policy discourse. The Council of Europe's convention of human rights and Biomedicine (the 'Oviedo Convention'), which entered into force in 1999, also formulates an approach based on the 'primacy of the human being', stipulating that: 'The interests and welfare of the human being shall prevail over the sole interest of society or science' (Council of Europe, 1997, article 2).

As such, when finalising this book in 2021, the 'human-centric' approach was a common term not only in theory but also in public discourse. Still, the said approach was presented in the policy discourse with no common conceptualisation other than an emphasis on the special role and status of people and the human being. As a stand-alone concept this could therefore mean many things, and as a matter of fact it also did in these policy discourses.[1] Some of these meanings could even be said to be ethically problematic, as also indicated by the philosopher Mark Coeckelbergh: 'A human-centric approach is at least nonobvious, if not problematic, in light of philosophical discussions about the environment and other living beings' (Coeckelbergh, 2020, p. 184).

Here, I therefore want to offer an explanation of what I prefer to refer to as a human approach that, yes indeed, takes its point of departure in human nature. However, it does not prioritise the wellbeing of the individual human being only. Rather it emphasises the role of the human as an ethical being with a corresponding ethical responsibility for not only the human living being but also for life and being in general (I will return to this claim with reference to Henri Bergson's concept of 'human morality', 1932/1977). We could also consider this the conceptual foundation of one particular human-centric approach, specifically advocated in the components of the European AI agenda and other governmental, civil society and technical community responses concerned with the ethics foundations of AI innovation and adoption. Let us here explore in more detail what I mean by this.

The human-centric approach was, in the European AI agenda, first and foremost framed in a European fundamental rights framework and with reference to a 'human-in-command' or 'human-in-the-loop' approach to the development of AI that supports and enhances human agency and decision-making. The International Outreach for Human-Centric AI Initiative set in motion in 2020 by the European Commission to engage globally, for example, described human-centric AI as an approach that ensures that AI works for people and protects fundamental rights. The human-centric approach was also explicated in technology and engineering standards aimed at designing human agency in data technologies, such as some of the IEEE P7000s standards or in the MyData movement that I have described in previous chapters. In this way, we may associate the human interest in the data of AI in practical terms with the involvement of human actors in the very data design, use and implementation of AI. In the HLEG on AI's ethics guidelines, the 'human-centric approach' is, by way of example, spelled out with particular attention to the interests of the individual human being, as well as the 'human-in-command' and 'human agency and oversight' components in the design and conditions for the development of AI. However, we can also trace the human-centric approach in more macro-societal calls for action, such as Pasquale's (2015) counter description to the 'black box society', the 'intelligible society', where decision-making processes are always intelligible to all humans involved on a technical and organisational, as well as a societal level. The human-centric approach is here actualised in what he refers to as 'humanizing processes' (Pasquale, 2015, p. 198), such as concrete legal frameworks which, for example, require the establishment of company and policymaking practices that embed 'human judgment' in decision-making processes (Pasquale, 2015, p. 197). A 're-humanization process' is also what the Danish privacy activist, Emma Holten, called her counter-activity to the dehumanising spread of her naked pictures online without her consent. As described earlier, she took control of the oppressive objectification of her body turning it into a female subject with a set of new photos that she shared online (Holten, 1 September 2014)

These are various proposals for a human approach to the Big Data Society centred around the value and promotion of human involvement and agency in legal, social, individual and technological processes. However, there is an extra layer of reflection to this. In the first part of the book (Chapters 1 and 2), I positioned a data ethics of power in the context of a recent data (re)evolution of the information society; that is, the evolution of the Big Data Society. A prevalent characteristic of the Big Data Society is that it is dictated by a transformation of all things into data formats ('datafication') in order to 'quantify the world' (Mayer-Schönberger & Cukier, 2013, p. 79). I also argued that the sociotechnical infrastructures of the Big Data Society, the BDSTIs and AISTIs, are not the manifestations of an arbitrary evolution, but can be viewed as expressions of societal negotiations between different cultures, interests and, at their very core, also worldviews and ontologies of the status and capacities of the human being and the role of data technology in society. With a postmodernist perspective we might even consider the sociotechnical infrastructures of the Big Data Society the materialisation of a prevailing ideology of the scientific practices of modernity to command nature and living things (Harvey, 1990; Jameson, 1991; Bauman, 1995; Edwards, 2002; Bauman & Lyon, 2013). The critical infrastructures of the Big Data Society can therefore be described as modernity embodied in what Paul N. Edwards describes as a 'lived reality' (Edwards, 2002, p. 191) of control and order: 'To live within the multiple, interlocking infrastructures of modern societies is to know one's place in gigantic systems that both enable and restrain us' (Edwards, 2002, p. 191).

With a human approach this can be identified as an ethical problem per se. Gilles Deleuze famously described over-coded 'Societies of Control' (Deleuze, 1992), which reduce people ('dividuals') to a code marking their access and locking their bodies in specific positions (Deleuze, 1992, p. 5). In other words, the human approach also voices the concerns of a postmodernist movement with the constraints of the practices of control embedded in the technological infrastructures of modernity that reduce the value of the human being (Frohmann, 2007, p. 63). As we saw previously, this was also a core critique carried forward in the field of surveillance studies (Lyon, 1994, 2001, 2010, 2014, 2018) as well as by Spiekerman et al. (2017) in their 'Anti-Transhumanist Manifesto', which directly opposes a vision of humans as merely information objects not different from other information objects (nonhuman agents); a vision which they, among others, describe as 'an expression of the desire to control through calculation' (Spiekerman et al., 2017, p. 2). In this way, we can also think of the human approach of a data ethics of power as a critical reflection on the power of technological progress and the sociotechnical systems we build and imagine.

However, to do so we need to first address the type of being-in-the-world that a data ethics of power presumes. I propose Henri Bergson's process ontology (Bergson, 1907/2001) and 'human morality' (1932/1977) as the first formative components of a data ethics of power. We can use Bergson's depiction of a human morality to approach a dynamic life in process, as opposed to a human life that is immobilised in data systems of social meaning making and representation. Moreover, Bergson raised an important critique of utilitarian approaches to the living, and therefore, I argue that his critique can also be used to illustrate the limits of the intellectual capacities that AISTIs possess, that is, as a type of intellect that may only reproduce syntax, but never semantics (Searle, 1980, 1997). With this perspective, the human approach of a data ethics of power is first and foremost an acknowledgement of the specific ethical potentials and responsibilities of humans that are very different from the intellectual potentials replicated in the autonomous moral agency of AISTIs. All in all, I want to provide a foundational understanding of the ethical capacities of humans and 'nonhuman' technological agents, such as AI, that make up our sociotechnical environment, and to propose an objective for data ethical action: to create the conditions of an open society for humanity. Bergson's concept of love without an interest is here the idea that will bring the action forward, recognising that this term is one that is shared in multiple cultures to signify a non-exclusive affection for all.

Henri Bergson's Human Approach

Henri Bergson's process ontology is one that essentially resists rationalist representations of reality as realities per se. They are representations that utilise the real for our own purposes, which is an ethical problem in itself. Bergson raised his key concerns with the limits of an intellect solely guided by a utilitarian approach to society in the early 20th century in the context of World War I and II. This is relevant to know, as his concerns also addressed and were informed by the experience of the severe real-life human implications of a particular approach to the living. Scientific and technical innovation was in war time invested and shaped by the conditions of war and the interests of enemies and allies, and he had seen some of the devastating human effects of scientific progress. In his last book, *The Two Sources of Morality and Religion*, published in 1932, Bergson argues that morality can only be set aside like this during a crisis, such as times of war, because it is practiced as a social moral obligation, not lived as a human morality (Bergson, 1932/1977). I will here illustrate what this means.

Most famously, Bergson illustrated his critique with reference to the human conceptualisation of time. Time is a human invention, he stated, or it is 'nothing' (Bergson, 1907/2001). Humans have created the mechanical structure of clock time to measure, segment and organise the time of the individual to function in society (Bergson, 1889/2004). However, clock time is not real time. It is the representation of the evolving time that we can only perceive while living it. Or phrased differently, we have two options for approaching life and reality: one is to approach it from the outside, rationally, with our own ready-made concepts, while the other is to experience reality in its 'creative evolution', as 'duration' (Bergson, 1907/2001, 1889/2004). The latter is what Bergson argued can only be achieved with human intuition. While indeed useful in science to act on material things ('matter'), the utilitarian intellect also, he argued, provides little room for a living, moving reality (Bergson, 1907/2001, 1896/1991, 1889/2004) and thus also an 'open' and 'inclusive' society (Bergson, 1932/1977). As he put it, 'In vain we force the living into this or that one of our moulds. All the moulds crack. They are too narrow, above all too rigid, for what we try to put into them' (Bergson, 1907/2001 p. viii).

Bergson also illustrates how the utilitarian intellect constitutes a type of morality. In *The Two Sources of Morality and Religion*, he proposes that there are two options for morality: a 'social morality' that takes form like our invented clock time, and a 'human morality' that takes form as the dynamic and mobile human time (Bergson, 1932/1977, p. 35–6). Social morality is expressed as a moral obligation that can be applied, but also set aside in, for example, moments of crisis. One might therefore also argue that it is prone to an interest-driven ethics. It can be used for purposes that serve specific interests. The utilitarian intellect, however, cannot produce the kind of 'human morality' that constitutes a way of being in the world that we do not put aside or apply when needed, but express as a 'style' or 'way of life' (Bergson, 1932/1977; Deleuze, 1986; Lefebvre, 2013). As such, Bergson also advocated a different ethical approach, a 'human morality' based on what he describes as *love* (I will return to this concept later). This is also what I present as the 'human approach' to ensure the open inclusive society that does not prioritise the human being per se, but rather a 'human morality' and an open unconditional love.

Henri Bergson's Process Ontology

Henri Bergson's process ontology is as complex as the reality he seeks to describe. To do this justice, I therefore also need to describe it in more detail here. Time is not only a metaphor in a Bergsonian ontology; it is also a philosophical approach and a method. Accordingly, the two types of time that I referred to before, or what Bergson refers to as 'multiplicities', also each correspond to a philosophical approach (Bergson, 1889/2004, 1907/2001).

One type of 'multiplicity' is abstract. It represents only spatialised time, the measure of time, not the temporal reality of time. It is a homogeneous time that is divided quantitatively, and accordingly changes in degree (changes spatial magnitude) when divided. Again, if we use the time metaphor, a clock represents this type of multiplicity. Clock time is not continuous but is divided into the instants 1 to 12; a sequence that is then repeated again. It is, as Bergson argues, a false continuum of time because it is abstracted from actual movement (duration) and is determined by spatial quantities (it increases every hour). It is time in a spatial form, which fragments duration into 'moments' and does not take account of what happens in the 'intervals' (Bergson, 1907/2001, p. 21).

The other type of 'multiplicity' is 'duration'/'real time'. Duration consists of multiple times that extend into each other like a 'flux of fleeting shades merging into each other' (Bergson, 1907/2001, p. 3). Duration is the 'whole' constituted by many different rhythms, of which human consciousness is just one of many. Duration is a heterogeneous multiplicity, and similar to Albert Einstein's relative time (but not the same; Einstein and Bergson were contemporary intellectuals agreeing and disagreeing throughout their lifetimes – Einstein disagreeing more with Bergson than the other way around). It is divided qualitatively; that is, when divided the whole of the 'moving zone' changes simultaneously (changes in kind, intensity). This kind of 'multiplicity' embodies what Bergson prefers to call the 'real' continuum of time; time in its temporal form, its 'continuous' form (Bergson, 1907/2001, 1889/2004).

The spatial segmentation of 'real time' (representation) into external homogeneous structures ('closed sets'), Bergson argues, is a utilisation of the real. Nature and the living are controlled and utilised for practical purposes. He reminds us that this homogenous structure is in fact not the 'real', but rather an objectification of the real. As such, the homogeneous multiplicity is an 'impure' continuum of time created by human reason, because it implies a stable universe; a state of 'being' where the 'whole' is given from the point of departure and each unit is aimed towards a predetermined point of closure. Conversely, reality is in truth a 'moving zone' (Bergson, 1907/2001, p. 3) or an 'aggregate of images' (Bergson, 1896/1991, p. 18) in which the human agent ('image') is one of many equally privileged agents ('image'). We might here also refer to Gilles Deleuze and Felix Guatarri's (1980/2004) more famous and quoted interpretation of reality as 'the field of immanence' (Deleuze &

Guattari, 1980/2004). The 'field of immanence' has no point of departure and no ending, and thus, it is continuous and open:

> not internal to the self, but neither does it come from an external self or nonself. Rather, it is the absolute Outside that knows no Selves because interior and exterior are equally part of the immanence of which they fused. (Deleuze & Guattari, 1980/2004, p.173)

In this reality ('aggregate of images' or 'field of immanence'), human conduct is, according to Bergson, limited when confined by a utilitarian intellectual approach. The utilitarian intellect will only grasp its own possible action upon other objects (Bergson, 1896/1991, p. 21) and what is apprehended is solely the 'best illuminated point of a moving zone' (Bergson, 1907/2001, p. 3). It is a type of intellect that is shaped and limited, as described previously, by the aim to utilise the real (Bergson, 1903/1999). However, humans also have other potentials; we are only limited if considering this type of intellect our only available capacity:

> all doctrines that deny to our intelligence the power of attaining the absolute. But because we fail to reconstruct the living reality with stiff ready-made concepts, it does not follow that we cannot grasp it in some other way. (Bergson, 1903/1999, p. 51)

The human mind is a composite of 'intellect' and 'intuition'. Intuition we can associate with the human's ability to 'think movement' (Bergson, 1907/2001, p. 318), our situated experience informed by a qualitative time or human memory (Bergson, 1896/1991) and more than any agent within a moving reality, the human thus has a potential for accessing the real time of reality, 'duration', as only humans have the potential to perceive duration with intuition. We can illustrate these two different types of intellectual capacity by comparing the 'creative skills' of an artificially intelligent system with only 'intellect' capacities and a human being with 'intuition' capacities. While AI software can be trained by processing the data of 346 Rembrandt paintings to successfully create a unique 3D-printed image that looks like a Rembrandt painting, perhaps even much better than a human reproduction, it could not do so without Rembrandt.[2] A human could not either. However, what a human could do is to produce her own painting, uniquely positioned in time and space.

Here, it also makes sense to distinguish between the application of the two very different capacities when applied in human dynamic environments, as there is a fundamental tension between them. Again, an AI system's approach is a good example. In December 2019, the AI company BlueDot Inc. demonstrated the great potential of AI big data predictive analysis when it raised an early alarm regarding a looming pandemic after having applied AI analysis to news reports and airline ticket data. Yet, in 2020 the big promise of predictive AI models like this was challenged by a human environment and behaviours in a fundamentally altered and unpredictable shape due to lockdowns worldwide and a global crisis (McLeod, 14 August 2020). An AI system assumes an ontology of immobility, or in other words, a predictable reality of things, including human environments; but if we take for granted Bergson's process ontology, the very properties of human environments are unpredictable and mobile. The historical training data on human behaviour that had been shaping AI predictions up until late 2019 simply could not deal with a present 2020 and a future beyond with unpredictable properties.

These different forms of human dynamic qualities (the very unpredictability of the human 'critical cultural moments') are challenged by the pervasiveness and potential normativity of AI systems. Take, for example, AI tools developed to support a judge's decisions. They process case law and then present a sum decision. A judge can use an AI tool like this to inform his or her own decisions. But we could also imagine that AI tools such as these become judicial normative AISTIs that privilege the quantitative AI analysis of past case law decisions over the qualitative situated judgement of the individual judge, and in this way locking, as one Council of Europe committee charter describes it 'his future choice into the mass of these "precedents"' (CEPEJ, 2018, p. 67). Phrased differently, but with a similar concern, the professor of political geography, Louise Amoore, introduces 'doubt' as the most human component of an ethical decision-making process, thus challenging the solidity of 'doubtless' decisions that are the result of machine learning processes weighing potential futures against each other and making room only for one probability:

> With contemporary machine learning algorithms, doubt becomes transformed into malleable arrangement of weighted probabilities. Though this arrangement of probabilities contains within it a multiplicity of doubts in the model, the algorithm nonetheless condenses this multiplicity to a single output. A decision is placed beyond doubt. (Amoore, 2020, p. 134)

Returning to the human approach of a data ethics of power, we can now qualify the previous proposition that this is, above all, an acknowledgement of the essential value of humans as ethical agents in a sociotechnical environment. In fact, the replication of a utilitarian intellect in nonhuman intelligent agents is a core ethical problem that a data ethics of power addresses, and here, we can also challenge the idea that 'intelligent' nonhuman moral agents can also be ethical agents. In fact, we can argue with Bergson that ontologically speaking they are not 'ethical beings'. Indeed, it is evident that a technical system that gains its 'intelligence' (learns, remembers and evolves) via data/the spatialisation of real time, can only possess one type of human intelligence, which is the one represented by what Bergson refers to as 'clock time'. That is, a data system is always spatialised time, taken out of its temporal context and immobilised to be utilised for the purpose of the system. This is why the first component of a data ethics of power is also a recognition of data ethics as a human responsibility.

Intuition as Method

With Bergson's process ontology of existence as movement, we are also presented with a method for a data ethics of power. To Bergson, the ideal philosophical action is 'to think movement' (Bergson, 1907/2001, p. 318), to become one with the object of analysis; that is, to constantly renegotiate stable meanings. As a form of perception, intuition has here a very specific status (Bergson, 1896/1991, p. 66, p. 185, p. 183). In his book *Bergsonism* (1966/1991), Gilles Deleuze refers to this as 'Intuition as Method'. This is an approach always moving towards an undefined point in the future as a state of 'becoming'. It is also the approach that enables what I refer to as the 'ethical agent' to place herself within the qualitative and temporal context of a critical problem or ethical dilemma and consider the conditions of the problem. In fact, it enables the ethical agent to not only find existing problems, but to also pose new problems. The ethical agent is, for that very reason, also ideally a free ethical agent.

Therefore, a starting point for a data ethics of power is to identify and reveal problems and solutions that may be covered up by stable systems of meaning production (or what I have previously referred to as dominant cultural systems or orders). As Deleuze describes it, solutions and problems are inseparable from the systems in which they exist, and this is also why they are not easily detectable. A first step is therefore to uncover 'the conditions of experience' (Deleuze, 1966/1991, p. 23) (or to use Stuart Hall's term 'conditions of percep-tion') that have shaped already posed problems and solutions, or what Deleuze also describes as the problem's 'means and terms of stating it' (Deleuze, 1966/1991, p. 15). In fact, this also includes uncovering 'false problems' that within one (dominant) cultural system of meaning production might be per-ceived as disorder, but in another is actually a cultural order in its own right. To illustrate this, we could think of the act of cleaning data of an information processing technological system according to a specific classification system. This might be done in an 'orderly' manner, strictly adhering to the rules of this classification system, but at the same time we may discover that the very mode of ordering, the classification system, is a critical problem in and of itself. The system might have methods for classification that are biased by represent-ing only one dominant group, causing an ethical problem, for example, for a minority group (as also illustrated by Bowker & Star, 2000). Here, it is the very order of the system that is the problem, and the 'disorder' (an alternative cultural system) that is the solution. Accordingly, we may argue that as a rule all data ethics problems are unique, just as all data ethics 'solutions' are unique – but also that they are uniquely interrelated as components of cultural systems of meaning making.

Deleuze provides a modus operandi that we can also use for a data ethics of power. We do not start with the solution. We go back to the problem and consider how it is 'made', how it is 'set up'. The very problem in itself is the expression of specific power dynamics that will guide the solutions we propose and in which we engage, so to speak, our ethical agency. Deleuze states that 'True freedom lies in a power to decide, to constitute the problems themselves' (Deleuze, 1966/1991, p. 15). What kind of social reality does the problem posed present? Who has an interest in solving this specific problem? To solve a problem, we need to find the problem, to invent it – which includes understanding how the original problems were stated – and to 'uncover' falsely stated problems (Deleuze, 1966/1991, p. 15–19). Only then can we consider solutions.

By way of illustration, privacy may be considered a problem in data technology and business innovation if framed by a 'big data mindset' (Mayer-Schönberger & Cukier, 2013) within a surveillance capitalist economy, and technical solutions may be formulated accordingly. Thus, in this view of the problems of the Big Data Society, we may justify the development of big data systems with tracking by default technical components and with little privacy protection and safeguards. However, if we go back to the formulation of this 'privacy as obstacle problem', we will discover that privacy is in fact not a problem, and it might even be a solution in itself. We can, for example, consider privacy a type of data technology and business innovation (Hasselbalch, 2013 B; Hasselbalch & Tranberg, 2016). Deleuze also describes the Bergsonian method as a 'struggle against illusion' (Deleuze, 1966/1991, p. 21), a type of discovery of the real by uncovering representations of the real, which is equated with the conditions of experience. This can only be done qualitatively, which includes the recognition of our own position in space. We take up a 'volume in space' with our very bodies and we similarly fill time with memory that links the different instants we perceive (Deleuze, 1966/1991, p. 25). In other words, our position and immersion in what we want to understand, study or solve, is simultaneously a strength and a weakness. It conditions our experience, but our human intuition also empowers us to uncover these very conditions.

Intuition as method constitutes a temporal approach to a temporal reality; a dynamic process in constant evolution. Thus, we may also argue that a data ethics of power does not have a material form – it is not a guideline, a set of principles, a law, an initiative, a manual – it is processual, and for that reason it has, first and foremost, temporality. This is a crucial proposition, as socio-technical systems are also constituted in time. They have, as I have illustrated in the first part of the book with reference to Hughes (1983, 1987) and Moor (1985), patterns that can be discerned on macro-scales of time. That is, when considering a specific data technology design, for example, we do not only address it as a type of occupation of space with particular properties, such as a set of fixed values ('good' or 'bad') that we can equally instil by design with another set of fixed 'good' values. We consider it a sociotechnical process, imagined, built, adopted, governed, reinvented and so on.

In summary, with the human approach of a data ethics of power, we do not only ask to *what* material thing do we apply data ethics (for example, to which technology design), we also need to ask *when* is data ethics possible? As I have illustrated throughout this book, data ethics spaces of negotiation are possible in critical cultural moments of social controversy, 'when all the moulds crack' (Bergson, 1907/2001, p. viii), the moments of ethical reflection where cultures clash and we make implicit values and interests explicit. These moments, as I have previously argued, are the most *human* moments, and I assert this again now with even greater force. That is, essentially the critical cultural moments are the entire point of a human approach that seeks to ensure that they do occur in the design, adoption and governance of sociotechnical development.

The conditioning of human moments is also why a data ethics of power is crucial in the context of BDSTIs and AISTIs. A core ethical problem unique to the sociotechnical infrastructures of the Big Data Society is the immobilisation of human culture and, accordingly, of critical cultural moments in big data systems. In other words, how do we support human critical cultural moments in an opaque sociotechnical development, in which cultural controversy and human interpretation are reduced to an automatic big data process? This is an ethical problem that we have to address with a data ethics of power.

Thus, when I claim that the human approach is associated with cultural processes and ways of being in the world, what I mean is that it concerns a human *infrastructure* of power. It wants to enhance the critical agency and ethical responsibility of humans. Take the example of the UK students who, in 2020, went to the streets to protest against the algorithm employed by the UK exam board, Ofqual, to generate their grades when exams were suspended due to the COVID-19 pandemic. Instead of the teacher's dynamic situated assessment of each individual student, they were graded by an algorithm that weighted in their school's historical performance. The result was that the grades of students from large state schools plummeted, while the grades of those attending smaller fee-paying schools increased (Hern, 14 August 2020). Now, one thing was the students protests – we all remember the images of those – but think about the moments that came before those events. One individual student, Laura Hodgson, described the moment when she received her lower grade as follows: 'I logged on at 8 am and saw the Cs and just started sobbing for about an hour' (Gill, 13 August 2020). We can think of this very critical meeting between Laura and a predictive algorithm as crucially important: the controversial moment where the situated experience of a human being clashes with that of a predictive algorithm introduced as an AISTI. Moments like these are the critical cultural moments, when powers become visible and social controversy take centre stage. These moments, as I have argued, are also the most human moments that we aim to preserve in very tangible ways with a data ethics of power. When the situated experience of a human, human memory and intuition, are sparked. When human critical agency is fuelled.

Data Ethics as a Whole Way of Life and Culture

I have so far described culture as a set of conceptual systems of meaning making that bring together communities with shared conceptual frameworks and resources. Cultural systems are also active systems in the sense that they have specific priorities, goals and ways of organising the world that are actively imposed in society when practiced by engineers, for example, and represented in material things such as our technological systems. Crucially, culture is constructed and invested with interests and thus a dominant culture is only one view of the world.

The field of cultural studies was founded on a critique of such stable dominant cultural systems. Raymond Williams criticised traditional elitist definitions of culture that presume a stable social reality constituted by enduring values, prescribed meanings and states of being: 'It is stupid and arrogant to suppose that any of these meanings can in any way be prescribed; they are made by living, made and remade, in ways we cannot know in advance.' He continued, with specific reference to culture in England, though with a much broader meaning:

> the only thing we can say about culture … is that all channels of expression and communication should be cleared and open, so that the whole actual life, that we cannot know in advance, that we can know only in part even while it is being lived, may be brought to consciousness and meaning. (Williams, 1958/1993, p. 10)

According to Williams this 'whole actual life' (Raymond William's 'real') can only be made meaningful in a society constituted by creative open systems of knowledge. Culture is many, and importantly, these cultures are not just one stable way of life, they are not only extraordinary (elitist), but 'ordinary' (Williams, 1958/1993, p. 6); that is, whole ways of life. Thus, he confronts dominant cultural systems with a conceptualisation of culture as something that may be challenged and rebelled against with alternative cultural systems of meaning making. Here, we may relate the Bergsonian process ontology of a data ethics of power to William's conception of a dynamic, creative and, crucially, inclusive culture. If anything, culture is not a given, it does not have a stable meaning; it is a state of becoming, in negotiation and contestation. Exactly for this reason, I argue that the cultural component of BDSTIs and AISTIs is a specific concern of a data ethics of power.

A data ethics of power considers technology a cultural product and techno-logical practice embedded in socially ordering cultural systems of meaning making. In sociotechnical systems such as BDSTIs and AISTIs, culture – as we also saw in Chapter 4 – is actively practiced and lived by individuals, and cultural codes are taken for granted as stable frameworks of meaning, sus-taining power for some, while repressing the freedom and agency of others. Therefore, we may also approach the very cultural systems of data technology and technological practice as data ethical problems that we should seek to solve.

Here, as I have illustrated, the very act of cultural criticism is particularly challenged by increasingly autonomous moral agencies where dominant cul-tural classifications of the world are actively reproduced and activated. This is why a data ethics of power first and foremost seeks to recreate and ensure the critical cultural moments in the development and adoption of AISTIs and BDSTIs, where culture is treated as creative, 'whole' and 'many', and where negotiations of cultural meaning are enabled, and voices of alternative cultures are brought forward and considered meaningful. As put forward in Chapter 1, one objective of a data ethics of power is to challenge established cultural systems of power to enable the voices of marginalised cultural experiences.

Now, how about the *ethics* of a data ethics of a power – the implementation? If culture is not one and stable, and if it therefore cannot be grasped sufficiently and represented honestly by ready-made cultural concepts, then how does an 'ethical culture' evolve? In a conversation with Didier Eribon, when asked about ethics, Deleuze answered: 'It's the styles of life involved in everything that make us this or that' (Deluze, 1986). Ethics is not the same as a moral obligation, not just the representation of the good and the bad; it is everything we do, our 'style' of living. Thus, ethics is not something that is just taught and learned. It is, like culture, lived and practiced. This is also what the philosopher Shannon Vallor describes as practices of moral self-cultivation or 'cultivating the moral self', which she links to shared cultures of moral values habitualised and practiced under specific favourable conditions (Vallor, 2016, p. 63).

In *The Two Sources of Morality and Religion*, Bergson similarly distin-guishes between the two approaches: 'social morality' and 'human morality' (Bergson, 1932/1977, p. 35–6). As presented previously, he describes social morality as one that is imposed as a moral obligation or duty in society. Therefore, we do not experience it as our own, and in this respect, it is a kind of morality that we can resist and set aside. Human morality, on the other hand, is a way of living, a type of ethical way of being in the world. It is part of us as human beings, not represented in symbolic systems but expressed intuitively as an emotion in practice:

> if the atmosphere of the emotion is there, if I have breathed it in, if it has entered my being, I shall act in accordance with it, uplifted by it; not from constraint or necessity, but by virtue of an inclination which I should not want to resist. (Bergson, 1932/1977, p. 48)

It is this human morality that I consider 'ethical action'; that is, the very agency of a data ethics of power. It is expressed in very subtle ways, in constant negotiation and contestation with a dynamic complex environment. It is articulated in the style of our actions and practices, the nuances of different 'technological styles' (Hughes, 1983) and of different styles of governance. Social morality, on the other hand, does not have a style: it is non-negotiable, inscribed in rules, and in machines.

On these grounds, we may also conclude here that the ethics of a data ethics of power is realised as a transformation of culture, of a way of life (Bergson, 1932/1977; Lefebvre, 2013). This is a particularly critical realisation if we want our human rights materialised in sociotechnical systems as more than just adherence to and compliance with rules (Lefebvre, 2013). Of course, this does not in any way imply that there should not be written laws and shared common frameworks in society. Not at all. All it means is that these very solutions to ethical problems are not data ethics. What a data ethics of power does is to ensure a critique, to enable negotiation and reflection that will always result in a compromise that in itself could pose new ethical problems. As such, data ethics never has a point of departure, nor does it have an end, but is constantly moving with its target.

Love and the Open Society

Now, we must return to the first formative question of a data ethics of power: Why is data ethics important? A human data ethics of power is important because it enables an open society. Bergson (1932/1977) describes two types of society: the 'open society' and the 'closed society'. The open society is also a type of open universal 'love' that has no interest but is universally directed at the whole of humanity (Bergson, 1932/1977). That is, the open society is characterised by a truly universal independent love. It does not have a specific object (or interest). Professor of politics and philosophy, Alexandre Lefebvre,

describes this in his analysis of Bergson's depiction of the purpose and function of human rights as a way of caring for and relating to ourselves:

> The open soul overflows with love, but it is not for anything in particular. Not for
> one's family or nation, certainly; but also, not for humanity or nature or gods or the
> universe. (Lefebvre, 2013, p. 92)

The open society is therefore also a just society, in that it does not depend on
any particular content and does not have a particular interest. To illustrate what
this means in practice, Lefebvre uses Jankelevitch's (1967/2005) example of
the man who walks down the street joyfully smiling at everyone he walks by,
but at the same time not at anyone in particular, and summarises open love as
follows:

> Love is a disposition or a mood. It is a way of being in the world, rather than a direct
> attachment to any particular thing in it. (Lefebvre, 2013, p. 93)

The closed society, on the other hand, has 'boundaries'; it is based on 'pref-
erence, exclusion, and closure' (Lefebvre, 2013, p. 88) and is expressed in
'authority, hierarchy, and immobility' (Lefebvre, 2013, p. 90). The closed
society is one in which love is always progressively directed towards an object
(the family, the nation, etc.), a kind of morality imposed as a duty within
a given society. It is represented and symbolic. Love and morality in the closed
society expresses a 'closed tendency', in that it is directed towards a specific
object. That is, love has an interest. It is dedicated to a specific group. Bergson
exemplifies the core problem with social morality in the example of war. As
described previously, he asks how human rights can be set aside in wartime,
and answers, only because human rights are realised as a moral obligation
towards a specific group formulating this exclusive tendency of moral obligations accordingly:

> Who can help seeing that social cohesion is largely due to the necessity for a com-
> munity to protect itself against others, and that it is primarily as against all other men
> that we love the men with whom we live? (Bergson, 1932/1977, p. 32)

Thus, we have here two different types of love that materialise in the world as
two different types of moral practice. One type of practice is adaptable, open,
inclusive and mobile, while the other is immobile, closed and exclusive. We
might also argue here, with reference to the way I use the concepts of ethical
agency and moral agency, that while the first can be designated as ethical
practice and the agency of unconditional love, the other is only moral practice
and the agency of obligation and prescription.

Lefebvre argues that it is in the very description of the open society in *The Two Sources of Morality and Religion* that human rights gain a central position, because they are 'the best-placed institution to overcome the closed tendency of society and morality' (Lefebvre, 2013, p. 83). Human rights are indeed considered universal, in that they are applicable to all of humanity, meaning that human beings in all societies must enjoy the same rights. However, Bergson worries that the true value of human rights is not realised, due to the way in which they are implemented in society in practice as a social morality. Lefebvre also maintains that human rights are not well expressed in moral obligation; in fact, the goal of human rights is in essence better aligned with a changing state of being, that is, it is a transformative goal. The function of human rights is to change the minds and personalities of individual human beings and also of the state. For example, they not only protect individuals but also do so by reviewing and reforming arbitrary national laws and practices. Moreover, human rights do not only take the form of an obligation and compliance with law; they are embedded in cultural practice (Lefebvre, 2013, p. 75–81).

To illustrate the difference between a human morality and a social morality, allow me to provide an example of the life of an ethical value and human right – privacy again – in the context of the evolution of BDSTIs in the 2010s in Europe. I will, as a point of departure, consider privacy a human-centric value that enables the kind of open society I have just described. This is an argument by itself that I will not address further, as it is explored and well argued elsewhere (for example, by Solove, 2001, 2008; Cohen, 2013; Hasselbalch & Tranberg, 2016 and 26 December 2016; Veliz, 2020, among others). In Europe, the right to privacy is also established as a legal right in the GDPR and the Charter of Fundamental Rights, which is embedded in member state laws and, as such, can be considered a well-established moral obligation in Europe. Nevertheless, one of the most profound challenges in history to privacy on a global scale (and the right to 'private and family life') was posed in the very brief history of the internet by evolving methods of surveillance, tracking and automated electronic systems of retention and correlation of personal data.

Following the mass surveillance revelations by Edward Snowden, the United Nations General Assembly in 2013 affirmed that the same rights that people have offline must also be protected online. This statement was based on the realisation that the power distribution and conditions of the Big Data Society were challenging not only the legal implementation of human rights, such as the right to privacy, but these new constellations of power were also enabling the questioning of the very justification of a human right such as the right to privacy. That is, in the short period in which the internet became a central part of global society, privacy, as a protection of and moral obligation towards the individual, was increasingly held up against other interests with strong arguments for setting privacy rights of individuals aside. While 'anonymity', and thus 'online privacy', was described in the 1990s as a unique opportunity offered by the internet to experiment with identity (Turkle, 1997) and gender (Haraway, 1985/2016), and challenge under its protection established forms of power and constituted market models (Vinge, 1981/2001), the right to privacy in the shape of online anonymity was also associated with things such as aiding and effecting identity theft, trolling (Donath, 1999), bullying (Kowalski et al., 2008), terrorism, and the illegal sharing of copyrighted material (Armstrong & Forde, 2003). At one point, the very concept of individual privacy was even deemed obsolete or 'no longer a social norm' (Johnson, 11 January 2010). As such, privacy, as the legal scholar Julie E. Cohen states, 'got a bad name for itself' (Cohen, 2013, p. 1904) and public discourse on privacy transformed increasingly legitimising arguments against the right to privacy as well as privacy-invasive business and state practices.

Furthermore, the very experience of individual privacy as a cultural compo-nent was influenced. In 2014, Verner Leth, Rikke Frank Jørgensen and I con-ducted a number of focus group studies among Danish youth regarding their use of social media. Although these young people did recognise in practice their own need for privacy with various forms of online identity management, we also found that they had simultaneously become resigned to the idea that to participate in social life with their peers, they would also have to accept signing off their right to privacy to the social media companies that were facilitating it (Jørgensen et al., 2013; Hasselbalch & Jørgensen, 2015). In fact, they had negotiated their own interest in privacy with that of a social media business model and agreed to set aside their right to privacy.

Nevertheless, in the slipstream of a sweeping data protection legal reform in Europe, a renewed public policy focus and public awareness of the privacy implications of BDSTIs were gaining traction, and in 2020 it was therefore also one of the core concerns in the debate in Europe on contact-tracing apps. Did this mean that debates on online privacy had now finally matured into a human morality, a way of life, or in other words, a data ethics culture that ensured it as a core value in European technology development, practice, adoption and experience? Or, as I propose, did it not just illustrate yet again the application of a social morality with specific interests? Why was privacy, for example, only an interest in the debates on contact-tracing apps while in other areas it seemed to be a lesser concern?

A data ethics of power would here urge us to look beyond the public debate on technical privacy to the subtler power struggles and interests behind it. In the case of contact-tracing apps in Europe, these were expressed in the wrestling of one power actor (European member states) with another (Apple and Google) which can be argued to have, in part, diverted the attention from other data ethics implications of not only contact-tracing apps but also other data-based technologies developed during the crisis (Hasselbalch & Tranberg, 2020). That is, with a data ethics of power, we can address the structural distribution of power in the Big Data Society and argue that the implementation of human rights, such as the right to privacy, in BDSTI and AISTI development in Europe, was still only expressed as a *moral* obligation in society – that is, one that could be applied or set aside according to interests and dominant power.

Based on Bergson's scepticism towards the realisation of human rights as a moral obligation, or as a social morality, we can also ask: Is this why privacy can be applied or set aside in the sociotechnical reality of the Big Data Society? I argue that indeed it is. In 2020, while data ethics time and again was affirmed as a moral obligation in strategy documents, principles and guidelines, it was still not a culture and a way of life, nor a technological style and practice, that could not be set aside or applied according to corresponding power interests.

Human Rights, Human Dignity and Love

Internationally we have a human rights system with origins in the 1948 UN Declaration of Human Rights, drafted by representatives from regions all over the world. It has various mechanisms in place for monitoring the compliance of state parties with their human rights obligations. Furthermore, the European Convention of Human Rights has been signed by 47 members states of the Council of Europe and it has a Court of Human Rights. In Europe, the Charter of Fundamental Rights enshrines rights for EU citizens into EU law. Mechanisms are in place to ensure the application of fundamental rights in EU institutions and members states when implementing EU law. Protection of personal data is here one fundamental right that is also spelled out in an extensive data protection regulatory framework. Over decades, human rights have been standardised and established in conventions and international agreements, institutionalised and implemented in law and practice worldwide. We may ask, then, why do we need the human approach and a data ethics of power when we have an international human rights system like this in place?

When 'data ethics' gained traction in academic, public policy and business discourses in the late 2010s, its role and function in society was often critically questioned, at times even referred to as 'ethics washing' (Wagner, 2018); as a mode of diverting the attention away from the real human rights implications of BDSTIs and AISTIs and the urgent need to proactively update the positive obligations of states to protect citizen rights online. The results, outcomes and requirements of various 'data ethics' initiatives were also considered unclear and impractical. Would they not water down existing human rights requirements and law and the need to renew obligations and mechanisms for ensuring their application in a rapidly evolving online sphere? The critiques were to a large extent also justified. Many dubious 'data ethics' and 'AI ethics' initiatives did indeed emerge in this period created by individual companies, organisations and states supporting various interests. In spite of this, I do still believe that there is a way to re-appropriate the role of data ethics in the context of human rights.

Spelling out a 'human approach' for a data ethics of power does not mean grasping at thin air. To begin with, it means that we build things, act and govern in a holistic manner to benefit the kind of open love that Bergson advocated in his work in the last century, dedicated to a human morality and to human rights. Love for humanity. Love in the most figurative, but also very literal and practical sense.

Henri Bergson was not only committed philosophically to the realisation of human rights but also in very practical ways. At the end of World War I, he worked closely with US President Woodrow Wilson's administration to establish the predecessor of the United Nations, the international peacekeeping organisation the League of Nations, and he was even appointed president of its international commission for intellectual cooperation (the predecessor of UNESCO). Furthermore, his writings were a great influence on the drafting of the Declaration on Human Rights – at least, according to the man who drafted them, John Humphrey himself (Curle, 2007).

Thus, the human approach of a data ethics of power, I claim, has everything to do with human rights. Yet, it has very little to do with the way in which human rights have been implemented in the online sphere so far. The very fact that human rights had to be reaffirmed to 'apply also online' (UN, 2013) illustrates how the 'social morality' of a human rights system, when implemented as a moral obligation only, can also be set aside in social processes of innovation, business and state conduct. We can here think of the evolution of the 'tracking-by-default' business model that developed untouched by human rights legal oversight and accountability over decades. This could only happen because data ethics initially did not take form as a human morality of cultural practices and human responsibilities of designers and entrepreneurs, but evolved as a 'social morality' only, separated from cultural processes of innovation and business development in legal compliance and 'check list' practices.

What we need to do now is to re-centre the human rights and data ethics debates in policy, business and public discourse in a common framework of reference. Here, I want to emphasise Bergson's concept of 'open love' as the foundation for a contemporary data ethics and even a contemporary human rights framework. Traditionally, 'human dignity' has been the common ground for the various expressions of human rights. As the preamble of the Declaration of Human Rights states: 'recognition of the inherent dignity and of the equal and inalienable rights of all members of the human family is the foundation of freedom, justice and peace in the world'.

However, as previously described, dignity is also a historically rooted concept. It was forcefully emphasised after World War II in response to a very real and present experience of the totalitarian regimes of power and the cruel undignified treatment of specific communities of human beings. The Declaration of Human Rights was drafted on this experience and the ethics of it was evidently a response to this specific historical experience.

What we need now is a common point of reference that not only responds to a dominant totalitarian regime of power but responds to a more general liquid form of power. It has also become increasingly and urgently evident that we need to expand our ethical agency beyond the care of human communities and the dignity of each individual human being, to include a love for the entire environment and ecosystem in which we exist. Furthermore, in the online sphere, we cannot only think of human rights as a shield against an external 'demon', as Bauman (2000) describes the ethical implications of an epoch, we need to think of an ethics of open love in practice, a type of human empowerment within a system that holds an internal 'demon' embedded in our everyday lives and infrastructures. Thus, what we need is a non-exclusive love for all living beings without prejudice as the foundation of a human approach. We need a human morality of human rights.

Is this type of love universal? Can it be extended across geographical and cultural borders? Throughout history global intentions have continually also brought about global tensions. There is always the possibility that love is appropriated and transformed in the context of dominant cultural and regional powers and interests. A data ethics of power must be particularly vigilant of the eternal tension between local cultures and new forms of colonialism expressed in sociotechnical infrastructural practices. It is a tension that has not been solved in practice, and neither do I claim to solve it here. Nevertheless, I believe that the human approach is one key to the global implementation of a data ethics of power in a culturally non-exclusive way. Human rights scholar Clinton Timothy Curle's (2007) reading of the Universal Declaration of Human Rights and John Humphrey's drafting of this in a Bergsonian perspective is particularly helpful here. He presents the idea of human fellowship as the basis of a form of universal ethics without 'a radical homogenization of cultures' (Curle, 2007, p. 154). As he says, it is about trying to meet a standard of excellence while also understanding the historical and cultural limits for achieving this (Curle, 2007, p. 23). This is a form of human rights that is responsive to changing contexts: human rights as processes of translation with an 'intention' (or to use a Bergsonian term an 'intuition') grounded in the lived experiences of humans. In this way, we should not only think of human rights as a system, but as a type of human experience that embraces pluralism. As

Curle also says, a Bergsonian human rights approach establishes the 'human rights project as an attempt to restore a sense of humanité to Modernity' (Curle, 2007, p. 153).

With this ontology of data ethics, human love and humanly empowering sociotechnical infrastructures, we can now move on to the second formative component of a data ethics of power, namely an action-oriented critical framework addressing the conditions of power in the age of big data. If ethics and morality are 'styles', 'cultural practices' and ways of life of each individual, then how can we claim that they can play any role in the context of governance in society? That is, is there an approach to help direct society in an 'ethical way' beyond imposing it as a moral obligation only?

2.   HOW CAN A DATA ETHICS OF POWER ACHIEVE THE 'GOOD SOCIETY'?

If the distribution of power in the big data cultures, technologies and societies that I have described in the previous parts truly are in disharmony with our being in the world, with the 'open society' (what I have just proposed is the 'good society' of a data ethics of power), what can we do in an attempt to achieve harmony? What role can a data ethics of power play in this?

In our book (Hasselbalch & Tranberg, 2016), we described data ethics as a social movement of change and action:

> Across the globe, we're seeing a data ethics paradigm shift take the shape of a social movement, a cultural shift and a technological and legal development that increasingly places the human at the centre. (Hasselbalch & Tranberg, 2016, p. 10)

Phrased differently, in the late 2010s a data ethics of power was expressed in society as a proactive agenda concerned with shifting societal power relations and interests in the age of big data.

In the previous chapters, I have described the data ethics of this agenda as a human approach concerned with making visible the power relations embedded in the Big Data Society and their conditions in order to point to design, business, policy, social and cultural processes that support human interest and power. Consequently, I have proposed that power and human ethical agency are the anchors to which we tie a data ethics of power. One fundamental concern of a data ethics of power is the power conditions of a human morality being reduced to a social morality in data systems and processes, which thus also inhibits an open society/inclusive love without an interest. Hence, we may also refer to data ethics as a rebellion against the reductive character of data systems and power, against exclusive power and, as follows, we can argue that data ethics is predominantly the voice of the minority, the underprivileged and disadvantaged of the sociotechnical data infrastructures.

A data ethics of power encompasses the role of technology, culture and society in shaping power structures of human agency and experience all together. It makes invisible power dynamics visible in their temporal qualitative micro-, meso- and macro-social and cultural contexts. This essentially means that a data ethics of power is constantly moving on different plateaus in an effort to encompass the one and the whole (the micro and the macro) at the same time.

Data Ethics of Power as Practice

A data ethics of power is a Bergsonian human morality in action. As such, it is also a form of governance. This statement might be perceived as a contradiction, as we have just learned that a human morality cannot just be *applied* (neither can it be set aside) but is *lived* as a process and practice. Nonetheless, I want to propose here that the very promotion of a human morality could play an essential role in the governance of sociotechnical change in the age of big data and AI.

A data ethics of power is first and foremost a human approach that ensures human ethical agency in, and responsibility of, data cultures. To this end, let me here repeat what has been said before about the concept of ethical governance in the first part of the book. I focus on the role of data ethics and specifically the human approach. This includes a reflection on the critical moments of sociotechnical development when values and interests are negotiated and explicated and 'spaces of negotiation' emerge. The technological momentum required for a large sociotechnical system to consolidate in society is not just an arbitrary composition of social, economic and cultural factors mixed together by an inexplicable will of nature (Hughes, 1983, 1987); it has 'human power' and thus it may be transformed into different modes of governance that will guide the direction, the values, knowledge, resources and skills that form the technological architecture of the system, its governance, adoption and reception in society. But how?

As described in Chapter 3, ethical governance is 'multi-actor', 'reflexive', 'open-ended' (Hoffman et al., 2017) and aims at setting in motion 'processes, procedures, cultures and values designed to ensure the highest standards of behaviour' (Winfield & Jirotka, 2018, p. 2). Building on this conception, I propose to include a 'data ethical governance' that addresses the complexity of the Big Data Society in particular and that works to ensure human ethical agency and responsibility in BDSTI and AISTI development. 'Data ethical governance' seeks and finds unconventional critical problems within the conditions and qualitative reality of the Big Data Society, and constructs and restates problems and their solutions accordingly. Critically, 'data ethical governance' questions conventional problems and solutions by inquiring how, when and why these problems and solutions are posed and created (who has an interest in the problems and solutions that we take for granted?) 'Data ethical governance' asks: Which composite of solutions will best address the context and conditions of ethical problems? And crucially: How do we ensure a data culture in which the status of the human being as an ethically responsible and critical agent is acknowledged and ensured? To be exact, a data ethics of power plays a crucial role as a human-centric frame of reference in the governance of BDSTI and AISTI development.

In conclusion, I propose that there are two key acts of a data ethics of power. These are necessary to achieve the critical 'data ethics spaces of negotiation' in which the values and interests are negotiated, problems identified and constructed, cultural compromises are laid bare and directions are re-centred on a human-centric distribution of power (Figure 1):

Make power and interests visible. One act is a disclosive and analytical process; a critical applied ethics concerned with data interests in the cultures and power dynamics of concrete data technologies and systems, data design and practices in companies and organisations, among engineers and users, and in politics. Interests and power dynamics can be discerned with a micro-level analysis of the very design of a data system; they can be examined with a meso-level analysis of, for instance, the construction of political strategies on AI and data or the constitution of multistakeholder groups, legal negotiation processes and so on; and they can be investigated with a macro-level analysis of, for example, cultural paradigm shifts, power dynamics and global cultural patterns. I have referred to several examples of critical applied ethics as this, such as investigations of data systems, critical data studies, studies of discourse, legal studies and surveillance studies.

Ensure 'critical cultural moments'. Another act is to ensure the human 'critical cultural moments' in sociotechnical development and adoption. I have previously argued that these 'critical cultural moments' have special *human* characteristics, meaning they are possible when human memory and intuition are privileged and provided time and space to tinker. Therefore, ensuring these moments is also essentially what the human approach is all about. An 'openness' to the human 'critical cultural moments' can be practically ensured on a micro-, meso-, and macro-scale of time in, for example, data design and processes, in institutional and company practices, and in the moments in between crisis and consolidation of a larger sociotechnical system in society.

In the end, with 'data ethical governance', much is up to more 'untraditional' forms of human governance; the developers and engineers of BDSTIs and AISTIs; the people in the companies and organisations; the people that educate and direct others, from primary school to university to the workplace;

the activists who reveal and fight against the injustice; the scientist who develops the methodology; those that deploy BDSTIs and AISTIs, procure them and adopt them. Nevertheless, there are also some critical 'data ethical governance' components of the traditional governance work of the policymakers in the Big Data Society. I have referred to these as 'legal frameworks that defend human powers' and 'bottom-up governance approaches shaping human involvement in AI systems'. I believe that in combination, all of these processes and practices can reshape the Big Data Society into a Human Society.

An Infrastructure of Human Empowerment

I have presented sociotechnical infrastructures of power, the AISTIs and BDSTIs. They are what a data ethics of power addresses, but not what it wants. What we want to ensure and develop in practice is an infrastructure of human empowerment. An infrastructure of human empowerment is one that is not just blindly trusted and relied on by humans. It is an infrastructure that constantly has to prove its worth to humans who, in this way, can take ethical responsibility for it. This is the only way we can think of a sociotechnical infrastructure as 'trustworthy'.

With Bergson, we can argue that human power rests with our situated experience informed by a qualitative time, 'memory', a qualitative heterogeneous multiplicity, or what we could also think of as a human potential to 'think movement', to perceive duration. What I want to say here, in more or less cryptic terms, is that a data ethics of power is not just *about* humans – it *is* human. Just like Bergson's duration is not about time (like clock time is), it *is* time.

Ontologically speaking, the duration of human environments resists the infrastructural clock time powers of AISTIs. I argue that they only do so explicitly and critically in moments of controversy – when the moulds crack, so to speak, and the clash between the different powers becomes visible. Controversy is the most human value: to resist and to reject the futures inscribed in our individual and collective presents. What does this actually mean in the context of what I have described as a competition between the three powers, AISTIs, BDSTIs and human power? Right now, AISTIs have increasingly powerful agencies immobilising the living by making it predictable in time. Just like Bergson's clock. We always know not only that the clock will strike 12, but exactly *when* it will strike. This is a comforting feeling, it empowers us to manage and coordinate in social environments, but it does not mean that everything must be known in advance. It does not mean that our futures are set. The unpredictability of our human lives and societies is what we want to preserve, because if injustice, wrongful treatment of minority groups and discrimination are set in stone, in the algorithms and their data systems that pervade society, we do want to have the power to resist the futures inscribed in their runes.

NOTES

1.   See, for example, excerpts regarding the 'human-centric' principle of the AI frameworks of the EU, Australia, Japan, Singapore and OECD: https://ai.bsa.org/global-ai-principles-framework-comparison/
2.   The 'Next Rembrandt' is a 3D-printed painting created as part of a Dutch advertising campaign with AI software. It was generated from the analysis of 346 Rembrandt paintings.

6.   Conclusion to *Data Ethics of Power*

'In vain we force the living into this or that one of our moulds. All the moulds crack.
They are too narrow, above all too rigid, for what we try to put into them.'
Henri Bergson, 1907

One day in 1940, the world-famous philosopher, Henri Bergson, went to register his data at a police station in Paris. As a person of Jewish descent, he was required to do so due to the Vichy government's new anti-Semitic laws that had just been introduced in France after its surrender to Germany. These laws, among other things, prevented a Jewish person from taking public office; being a member of the press, a student, a doctor, or a lawyer; and having a business. Because of his status as a renowned academic, Bergson had been offered to be excused from these laws by the Vichy government, but he had refused. At the police station, completing his police form, Bergson wrote: 'Academic. Philosopher. Nobel Prize winner. Jew.' (Martin, 1994/2014, Chapter 10).

Eighty years later, in 2020, in another part of the world, the United States, Robert Williams was arrested in front of his house and brought to a police station where he was held overnight for a crime he had not committed. In fact, he had been wrongfully arrested based on an erroneous biased match from a facial recognition system used by the police. Facial recognition systems like this had at that time been used by police forces in the US for more than two decades. Deployed for surveillance of specific communities and to identify people for prosecution, these systems had, time and again, also been exposed as reinforcing racial bias. Presented with a grainy picture of the identified criminal, a black man, like Williams himself, but clearly not him, his first reaction was to say: 'I hope you guys don't think that all black men look alike' (Hill, 3 August 2020; Williams, 24 June 2020)

The data systems that we create to make sense of, organise and control life and society have, throughout human history, always reinforced power dynamics – often with devastating consequences for the human life represented in and by these systems; however, they have also changed shape. Today, the transformation of all things into data as an effortless, costless and seamless extra – and most often invisible – layer of life and society is one variety, which I in this book have argued requires a particular reflection and awareness from us. The data systems in which Henri Bergson was registered in 1940 and Robert Williams in 2020 both clearly represented and reinforced ethnic bias in

159

society, and both systems constituted devastating ethical implications for the people to which they were applied. Nevertheless, there are subtle differences. For example, while Bergson did not choose the system, he chose the data; a tiny personal rebellion, but nevertheless his comment against a devastating data system of power. Williams' data, on the other hand, was chosen for him. In fact, he was not even aware that he had been registered in the data system that was now being used against him by the police.

We are today not challenged by a database and register of a dominant regime of power, we are submerged in sociotechnical data systems of power. This is why we need what I have referred to in this book as a data ethics of power – to make powers visible and create the human critical cultural moments that ensure data ethics spaces of negotiation in society.

I will go through the main propositions of the book here, but first I want to address one of the core reasons why I wanted to write it. This is also connected with issues of power, but more substantially related to the way in which we conceptualise and apply data ethics. As I have pointed out several times throughout the book, data ethics is not only about power – it also *is* power. Power for governments, companies, self-proclaimed experts and advisors, and even academic disciplines to point out the problems and their solutions, to set the priorities for what role data technologies should play in our human lives and in society. Often, this is done with voices that speak louder, with more force (more resources, political and public discourse power) and at the expense of other voices, and not the least experiences of data power. This is also the reason why data ethics – as a field of study, as an approach, as a concept – is in crisis: blamed (often rightfully so) for a lot of things, but most critically for being a sweet cover for the bad data practices of companies and governments. For being a form of 'ethics-washing' in technology practice and governance. Said in other words, we might say that data ethics is what everyone knows, yet, as I have tried to illustrate in this book, no one can really say it is theirs. This is also why the concept is prone to being taken hostage by various interests. That is, precisely because data ethics does not want to belong, because it resists ready-made concepts, it can also become anyone's declaration.

I ask here: Why? What if we decide to set data ethics free? To uproot the very conceptualisation of the term as the moral obligation of someone or something to solve a specific problem? In a way that data ethics truly becomes a method and practice for humans to critically challenge the power embedded in data technologies, their set priorities and restraints, and to find different problems and new solutions in the very conditions of the big data reality we live in. In this way, power is no longer just an arbitrary condition, it is the material we act upon. Let us explore now the conditions for, the application of and the very shape of such a data ethics of power by going over the three main parts of the book and then discussing what will hopefully follow.

I.   Power and Big Data

In the first part of the book, I examined power and big data (Chapters 1 and 2). I described sociotechnical infrastructures based on big data technologies, which I referred to as Big Data Socio-Technical Infrastructures (BDSTIs). In the 2010s they were increasingly representing and constituting global societies and environments as the mundane background against which social practice, social networking, identity construction, economy, culture and politics were conducted. They were in part institutionalised, in systems requirements standards for IT practices and in regulatory frameworks for data protection; they were invested with politics and human imagination about the challenges and opportunities of big data; and they were up for negotiation and contestation.

I explored the BDSTIs of the Big Data Society as a particular type of architecture of power for human agency and experience. BDSTIs are constituted as material global networks that enable data collection and access across geographic territories and jurisdictions, and as flows around which dominant societal functions are increasingly organised (Castells, 2010). Accordingly, to design and shape the infrastructural components of BDSTIs is also an essential form of power.

A data ethics of power addresses the new conditions of power of these technologically mediated configurations of space and time and their ethical implications. For example, BDSTIs' spatial organisation of power is created for and by new types of dominant 'managerial elites' (Castells, 2010). Traditional arbitrary surveillance powers of states are therefore additionally augmented by the powers of commercial actors that design BDSTIs to accumulate, track and access big data (Hayes, 2012; Pasquale, 2015; Powles, 2015–2018; Zuboff, 9 September 2014, 5 March 2016, 2019). As a result, surveillance powers of states and commercial actors alike are embedded in BDSTIs as a key property of their architecture and design (Haggerty & Ericson, 2000; Lyon, 2001, 2010, 2014, 2018; Hayes, 2012; Bauman & Lyon, 2013; Galic et al., 2017; Clarke, 2018).

In the first part of the book, I also presented sociotechnical infrastructures, like BDSTIs, as social spaces occupied by interests in constant negotiation (Lefebvre, 1974/1992; Harvey, 1990) and, accordingly, accompanying infrastructural practices as expressions of different political and cultural interests. In Europe, the idea of a 'European infrastructure' has generally been invested with the imagination and interest of an EU project that wants to enable the efficient workings of a union of collaborating member states. That is, infrastructural practices, such as engineering and design standards, construction, investment and regulation are defined as a strategic endeavour to produce a space that enables a European economic and social union and community. This political aspiration has also been invested in the idea of a European digital single market, translated in the 2010s into an aspiration to create European BDSTIs. The European BDSTIs are, first and foremost, defined here as differentiators on a global competitive digital market. Aspirations to compete in a global big data economy while preserving and protecting Europeans' fundamental rights were, in the late 2010s, reconciled in what has also been referred to as the European 'third way' (with a particular emphasis on the development of European BDSTIs with AI capabilities, what I refer to as AISTIs).

As such, the power of the big tech commercial elites of the global BDSTIs were increasingly challenged in Europe with concrete infrastructural practices, such as investment and policies supporting the development of European practitioner and user competences, science and research, technical data infrastructures and data pooling, as well as the development and possible implementation of legal frameworks to ensure the development of European BDSTIs and AISTIs.

When we explore the various infrastructural practices invested in the development of BDSTIs and AISTIs in the early 21st century, we see a moment of conflict and negotiation between the conception of a European cultural space, the BDSTIs and AISTIs of this space, and existing BDSTIs and AISTIs, which is increasingly defined as a 'foreign' space. In this first part of the book, I argued that a data ethics of power has a crucial role to play in critical moments such as this, as they lead to the cultural compromises or 'technological momentum' (Hughes, 1983, 1987) that a global sociotechnical system needs to change and evolve. They are also crucial to phases of innovation and development as they constitute the very transformation of the sociotechnical system that emerges out of a quest to solve critical problems of the system. Data ethics can therefore also be identified in institutions, companies, governments and intergovernmental organisations that either move along the natural state of affairs or establish initiatives and practices, namely 'spaces of negotiation', dedicated to the value and ethical reflection that must accompany legal compliance.

II. Power and AI

In the second part of the book, on power and AI (Chapters 3 and 4), I examined the history and special power characteristics of BDSTIs with AI capabilities – the AISTIs. AI has progressed from the rule-based expert systems encoded with the knowledge of human experts and applied in primarily human and physical environments to systems evolving and learning from big data in digital environments with increasingly autonomous decision-making agency and capabilities.

I also examined the ethical implications of AI (and thus the ethical implications of AI technologies' evolution into AISTIs in the late 2010s) and the ethical theories that address these specific implications. I narrowed these down to concerns with the relations between the distributed 'ethical' agency of humans and the 'moral' agency of AI; a complex that is increasingly shaping our ethical experiences in profound ways. I here consider an overarching theme in the AI applied ethics research field regarding the different degrees of human involvement in AI systems' design, adoption and consolidation in society. These concerns can also be traced back to ethical concerns regarding AI's threat to humanity and human control or, in contrast, the potential of autonomous AI to surpass human deficits. However, I do not adhere to either one or the other poles of that discussion. Instead, I propose a middle way. Accepting that AI can be a moral agent does not inevitably imply that we have to accept it as also an ethically responsible agent. That is, the question should not be whether machines should or can have human-level ethical agency and responsibility; rather, we need to focus on the ways in which we can ensure humans continue to be involved in a meaningful and responsible manner. For instance, we can in very concrete ways create new standards and laws for AI and robotics to ensure such human involvement and empowerment in AISTIs (Pasquale, 2015, 2018, 2020). Furthermore, I also propose that we consider the very cultural systems in which AI gains its meaning as an either uncontrollable autonomous moral agent or as a human data design in which human involvement is a key property.

For this reason, I argue that a core ethical concern of a data ethics of power is with AISTIs and BDSTIs' constitution as cultural normative systems with a type of social ordering, in which interests of dominant actors in society have the primary advantage. I have explored specifically the cultural systems that shape the practices of developers, scientists, lawmakers and users of AISTIs. Here, I examined more closely the interests in the data of AI as expressions of different data cultures, with reference to the cultural systems of meaning-making that shaped the AI technological momentum of the late 2010s. A concern of a data ethics of power is with technological change as a field of power negotiation between the interests of different technological cultures, their compromises, or the dominance of one cultural system over another. With my point of departure in Hughes' description of 'technological momentum' (1983, 1987), I have proposed adopting a helicopter analytical view on interests as a set of complex factors that come together in shared cultural knowledge frameworks and worldviews that cut across different stakeholder groups and communities. Based on these explorations and arguments, I propose that an ethical concern with AISTIs is when human critical negotiation and ethical agency (the 'critical cultural moments') are immobilised in AI moral agency.

III. Human Power and Data Ethics

In the first two principal parts of the book, I formulated a data ethics of power as ethics critically engaged with the negotiations between the powers and interests of different sociotechnical data cultures of BDSTIs and AISTIs, their conflicts and ethical compromises. BDSTIs and AISTIs constitute two forms of power that 'work' in different dimensions of human reality and society. While BDSTIs primarily act in space by transforming all into immobilised digital data, AISTIs also occupy time by acting on that data to actively shape the past and the present in the image of the future. Thus, as argued, a core concern of a data ethics of power is with these AISTIs' and BDSTIs' constitution as cultural systems of a type of social ordering in which interests of dominant actors in society are spatialised and immobilised, and thus, more difficult to be critical of and renegotiate.

In the final part of the book (Chapter 5), we looked at the special characteristics of human power and ethical agency. These are crucial to any engineered or non-engineered act of ethical governance of sociotechnical change as well as at the heart of a data ethics of power. A data ethics of power is ultimately concerned with the role of the human as an ethical being with a corresponding ethical responsibility for not only the human living being but also for life and being in general. We need this human approach as a guiding narrative for the governance of AISTIs and BDSTIs; that is, we need an approach that prioritises the human environment, human ethical agency and responsibility. As I showed in the last part of the book, I use this concept with much explicit devotion to human judgement, governance and critical situated experience, as opposed to the moral agency of technological artefacts that can only represent, reproduce and reinforce living things without experience and critical agency. To illustrate this with a very simple example from the realm of our ethical concerns with AI: while an AI agent does not negotiate its own bias, as it can only operate as if it was a 'complete' system, a human might not see or care about its own bias, but has the *potential* to see it, care about it, and be confronted with it in cultural critical moments.

In summary, human power needs specific spatial and temporal conditions to flourish, but in the early 21st Century in particular, human ethical agency is in constant negotiation with the moral agency of BDSTIs and AISTIs. They not only materialise power dynamics but they also challenge human ethical agency to negotiate and revolt against these powers in fundamental ways. Or said in other words, human critical ethical agency does not resemble, correspond with, live well, or flourish in the cracks of the BDSTIs' and AISTIs' sociotechnical structures of power. This is why we urgently need to act to preserve the human critical cultural moments and to create the conditions that will enable critical spaces of negotiation. We need to actively build alternative sociotechnical data infrastructures and systems that interact with human agency, power and ethics in a different way. We need applied data ethics in design and governance that ensures the involvement of human life, experience and critical agency in the very data design, governance, use and implementation of sociotechnical data systems.

What's Next?

During my research for this book, I stumbled upon the delineation of the design of a mechanical 'ethical governor' of autonomous weapons. This would be a technical component created to process the stream of data that shapes the agency of an AI warrior agent, a type of mechanical data control component built into the agent's data design to ensure that it behaved within a set of prescribed ethical boundaries (Arkin et al., 2009). The ethical governor would work by processing the data of the agent's fields of action and permit or forbid actions based on this. There would, for example, be the data on humans, essential buildings and meaningful cultural sites, which the agent's data design would transform into data streams of forbidden lethal action; then, there would be data on threats that could also include data on humans, as well as data threat scenarios that, in combination with other data on the ethical bounds of the ethical data governor, could be transformed into data torrents of permissible lethal action. This represents hundreds of years of military law and complex ethical boundary creation transformed into the data process of a data design.

Making ethical decisions on life, death, culture, bombs and lethal action is exceedingly complex, but ethical decision-making transformed into a data process appears to be less so. Today, this reduction of human complexity, I argue, is one of the key motivations for BDSTI and AISTI development and adoption: to make life, society and culture easier to handle; and to make those difficult ethical decisions we must make every day less cumbersome as individuals at home, work, school, and in our societies, in hospital, during elections, in the welfare system, in the justice system and during times of crisis, such as war or pandemics. The philosopher, Henri Bergson, who went to register at a police station in Paris in 1940 is crucial to this book. But so is the man, Robert Williams. In one split second the complexity of his entire being was reduced to a few correlated data points, he was submerged in a contemporary digital data power complex, and he was arrested by police officers relying on a facial recognition system's erroneous match between his face and that of a criminal.

We have to make thorny human decisions with ethical implications every day, and time and again we realise ourselves, or are told by other humans that we have made them poorly or in ethically problematic ways. However, the human self-conscious critique, which a data process does not have (as Louise Amoore (2020) puts it, an algorithm does not doubt itself), is also the very reason why we need to keep making these decisions ourselves. Think about an ethical evaluation of the most critical situation you can imagine, such as releasing a bomb during wartime, transformed into a data process without critical agency. It is terrifying. Nonetheless, that is precisely the kind of imagination concerning the reduction of complexity and human dilemmas of ethical decision-making that drives much of the development and adoption of BDSTIs and AISTIs today.

It is also this loss of human critical agency that I most fear will take form in our sociotechnical realities if we do not halt and redirect current BDSTI and AISTI developments. We do not seem to realise it, because cutting out the human does not mean cutting out moral agency (making a decision, even a moral one, is something a data process certainly can always be designed to do). However, in this process, as I have tried to illustrate in this book, we cut out the kind of critical ethical human agency that is fundamental to our democratic societies and their institutions. Human critical agency is what we remove from the very configurations of our sociotechnical spatial architectures and from our societal imagination, our norms and cultures.

Now, we have some core tasks ahead of us to steer the change we need. These will all go into the development and consolidation of a highly complex sociotechnical system that will be shaped by a multitude of cultural, economic and social factors; thus, I can of course only mention a few of the tasks here.

The first thing we need is a new imagination, a different technological culture or 'data culture' of human power, when building and adopting AISTIs and BDSTIs. We have to challenge the very yearning for the perfect society, the efficient society and even the just society that materialise in their design cultures. Humans, biology and societies are messy, unpredictable things. They are anything but perfect; often they are unjust, and definitely worthy of our ethical concerns and critiques. However, big data systems or AI will not change that. Only humans can make real changes by ensuring the conditions and structures of power that enable human critical agency. What we can do, however, is to imagine BDSTIs and AISTIs as potentially incredibly useful *tools* that can support human critical and ethical decision-making with evidence; for example, *apparatuses* that can make a scientific analysis stronger, or *instruments* that can indeed help us make many processes more efficient.

We need technical components of sociotechnical infrastructures that enhance the critical agency of individuals. We absolutely must ensure meaningful human involvement, which we can do by carving out a place for individual human experience and agency in the sociotechnical fabric of BDSTIs and AISTIs. The first place to start, which particularly Europe seems to be investing in at the beginning of the 2020s, is the development of a data infrastructure that by default respects people's privacy and empowers individuals. Today, such technical components of a data infrastructure go by many names (data trusts, personal data management systems, personal data stores, and so on) and they take on many forms. Much is left to be explored in terms of their basic functioning, interoperability and, not least, their legal framework. Here, we also need to build in by design meaningful 'human-in-command' components, and we need data systems in which critical functions and criteria can be explained.

Moreover, we need people who understand big data and AI systems well enough to engage with them in a meaningful manner when they use them, procure them, create laws about them and build them. We need sociotechnical data literacy and education of children, educators, students and policymakers.

Evidently, there is no way around our normative legal frameworks. We need to examine and update these to ensure the legal implementation of meaningful human control in the development and adoption of BDSTIs and AISTIs and to defend human expertise (Pasquale, 2020).

We also need more critical data studies that discern the data interests in the cultures and power dynamics of specific data design, practices and data politics. I have referred to several examples of brilliant investigative studies in these areas; however, most have been conducted in Western contexts, making visible their power dynamics and challenging the ethical problems and dilemmas of BDSTIs and AISTIs in these specific cultural spheres.

Crucially, a data ethics of power does not speak with only one voice and it certainly does not speak with the voice of the most powerful. Questions of power distribution, domination and advantages through big data sociotechnical systems are core components of a data ethics of power, but the experience of data surveillance and power is not homogeneous. The essential experiences of the people, communities and cultures that are most disadvantaged by the global big data structures of power needs a much stronger voice. In fact, a limitation of this book is also my Western/European perspective, speaking from a privileged socioeconomic position in a global environment of societies that move at different rhythms and are positioned with different advantages or disadvantages in the evolving big data sociotechnical systems of our age. This is why we urgently need more data ethics of power studies based on different cultural and socioeconomic experiences.

Last but not least, how about human 'ethical governors'? What if we designed independent (I mean truly independent in terms of not only finance and interests, but also in terms of their cultural conceptual frameworks of meaning making/imagination and narratives), multi-*expertise* (not multi-*stakeholder*)

ethics councils and bodies into the governance of AI and big data sociotechnical development? They would critically assess ethical dilemmas and value
clashes, working on the three scales of time and analytical level of abstraction
that I have considered in this book (the micro, meso and macro) inside organisations and businesses, outside as independent auditors, and with policymakers
when policy and law are negotiated. In terms of the power structure for the
independent governance of these, states have stakes in this that are too high,
and certainly the industry never asked for or intended to administer such
a function. Therefore, why not give civil society organisations the chance
now? Specifically, civil society and nongovernmental organisations that have
proven long-term independency and dedication to the human interest. Provide
these civil society agents, the representatives of love and human agency, with
a real governance role beyond the 'activist' role. Give them the resources to
compete, to professionalise, to be the 'ethical governors' of the age of Big
Data. Make the structural changes.

Bibliography

Acker, A., Clement, T. (2019) Data Cultures, Culture as Data - Special Issue of Cultural Analytics. Journal of Cultural Analytics. https://doi.org/10.22148/16.035

Adam, A. (2008) Ethics for Things. Ethics and Information Technology, 10 (2–3), 149–15. https://doi.org/10.1007/s10676-008-9169-3

Advocates for Accessibility (3 May 2020) Free Basics and Digital Colonialism in Africa, Global Digital Divide. https://www.globaldigitaldivide.com/free-basics/

Agger, B. (1992) Cultural Studies as Critical Theory. Falmer.

Aguerre, C. (2016) Agenda Building and the Internet: The Case of Intermediaries. Universidad de San Andrés. http://bibliotecadigital.tse.jus.br/xmlui/handle/bdtse/3263

AI Task Force and Agency for Digital Italy (2018) Artificial Intelligence at the service of the citizen. https://libro-bianco-ia.readthedocs.io/en/latest/

Albury, K., Burgess, J., Light, B., Race, K., Wilken, R. (2017) Data cultures of mobile dating and hook-up apps: Emerging issues for critical social science research. Big Data and Society. https://doi.org/10.1177/2053951717720950

AlgorithmWatch (2020) Automating Society Report 2020. Fabio C., Fischer, S., Kayser-Bril, N., Spielkamp, M. (eds.) AlgorithmWatch gGmbH

Allam, Z., Dhunny, Z.A. (2019) On Big Data, Artificial Intelligence and Smart Cities. Cities, 89 (June 2019), 80–91. https://doi.org/10.1016/j.cities.2019.01.032

Allen, C., Smit, I., Wallach, W. (2005) Artificial Morality: Top-down, Bottom-up, and Hybrid Approaches. Ethics and Information Technology, 7(3) (September 2005), 149–55. https://doi.org/10.1007/s10676-006-0004-4.

Alpaydin, E. (2016) Machine Learning. MIT Press.

Altman, A. (1983) Pragmatism and Applied Ethics. American Philosophical Quarterly, 20(2) (April 1983), 227–235.

Amoore, L. (2011) Data Derivatives on the Emergence of a Security Risk Calculus for Our Times. Theory, Culture and Society, 28(6), 24–43. https://doi.org/10.1177/0263276411417430

Amoore, L. (2020) Cloud Ethics Algorithms and the Attributes of Ourselves and Others. Duke University Press.

Anderson, M., Anderson, S.L. (eds.) (2011) Machine Ethics. Cambridge University Press.

Anderson, S.L. (2011) How Machines Might Help Us Achieve Breakthroughs in Ethical Theory and Inspire Us to Behave Better. In M. Anderson and S.L. Anderson (eds.), Machine Ethics (pp. 524–530). Cambridge University Press.

Angwin, J., Larson, J., Mattu, S., Kirchner, L. (23 May 2016) Machine Bias. Propublica. https://www.propublica.org/article/machine-bias-risk-assessments-in-criminal-sentencing

Arkin, R.C., Ulam, P., Duncan, B. (2009) An Ethical Governor for Constraining Lethal Action in an Autonomous System. Georgia Institute of Technology Atlanta Mobile Robot Lab, Technical Report GIT-GVU-09-02.

Armstrong, H.L., Forde, P.J. (2003) Internet anonymity practices in computer crime. Information Management and Computer Security 11(5), pp. 209–215. doi: http://dx .doi.org/10.1108/09685220310500117

Åsberg, C., Lykke, N. (2010) Feminist Technoscience Studies. European Journal of Women's Studies, 17(4) (November 2010), 299–305. https://doi.org/10.1177/ 1350506810377692.

Augusto, C., Morán, J., De La Riva, C., Tuya, J. (2019) Test-Driven Anonymization for Artificial Intelligence. 2019 IEEE International Conference on Artificial Intelligence Testing (AITest), Newark, CA, USA, 2019, 103–110.

Awad, E., Dsouza, S., Kim, R., Schulz, J., Henrich, J., Shariff, A., Bonnefon, J-F., Rahwan, I. (2018) The Moral Machine Experiment. Nature, 563, no. 7729 (November), 59–64. https://doi.org/10.1038/s41586-018-0637-6.

Baudrillard, J. (1990) Cool Memories: 1980–1985. Verso.

Bauman, Z. (1995) Life in Fragments: Essays in Postmodern Morality. Blackwell.

Bauman, Z. (2000) Liquid Modernity. Polity.

Bauman, Z. and Haugaard, M. (2008) Liquid modernity and power: A dialogue with Zygmunt Bauman. Journal of Power. https://doi.org/10.1080/17540290802227536

Bauman, Z., Lyon, D. (2013) Liquid Surveillance: A Conversation. Polity Press.

Barlow, J.P. (8 February 1996) A Declaration of Independence of Cyberspace. https:// www.eff.org/cyberspace-independence

Barthes, R. (1972) Mythologies. Selected and translated by A. Lavers. The Noonday Press. (Originally Published in French, 1957).

Barthes, R. (2000) Myth Today. In S. Sontag, A Roland Barthes Reader (pp. 93–149) Vintage. (Originally published 1982).

Bartoletti, I. (2020) An Artifical Revolution on Power, Politics and AI. The Indigo Press.

Beal I (i.e. Beall), A. (30 May 2018) In China, Alibaba's data-hungry AI is controlling (and watching) cities. Wired. https://www.wired.co.uk/article/alibaba-city-brain -artificial-intelligence-china-kuala-lumpur

Beck, U. (1993) Risk Society. Towards a New Modernity. London: SAGE Publications.

Beck, U. (2014) Incalculable Futures: World Risk Society and Its Social and Political Implications. In: U. Beck (ed.) Springer Briefs on Pioneers in Science and Practice, vol. 18. Springer, Cham.

Belli, L., Zingales, N. (2017) Platform Regulations. How Platforms are Regulated and How they Regulate us. FGV Direito Rio.

Ben-Shahar, O. (2019) Data Pollution. Journal of Legal Analysis, 11, 104–159. https:// doi.org/10.1093/jla/laz005

Bentham, J. (1787) Panoptikon: or, the Inspection-House. Thomas Byrne.

Bergson, H. (1977) Two Sources of Morality and Religion. Translated by A. Audra and C. Brereton. University of Notre Dame Press. (Originally published in French, 1932).

Bergson, H. (1991) Matter and Memory. Translated by N.M. Paul and W.S Palmer. Zone Books, Urzone. (Originally published in French 1896).

Bergson, H. (1999) An Introduction to Metaphysics. Translation T.E. Hulme. Hachett Publishing Company. (Originally published in French, 1903).

Bergson, H. (2001) Creative evolution. The Electronic Book Company Ltd. ProQuest Ebook Central. (Originally published in French 1907).

Bergson, H. (2004) Time and Free Will: An Essay on the Immediate Data of Consciousness. Taylor and Francis Group. ProQuest Ebook Central. (Originally published in French in 1889).

Bigiotti A., Navarra, A. (2019) Optimizing Automated Trading Systems. In T. Antipova, A. Rocha, (eds.) Digital Science. DSIC18 2018. Advances in Intelligent Systems and Computing, vol. 850. Springer, Cham. https://doi.org/10.1007/978-3-030-02351-5_30

Bijker, W.E. (1987) The Social Construction of Bakelite: Toward a Theory of Invention. In W.E., Bijker, T.P., Hughes, T. Pinch (eds.) The social construction of technological systems. MIT Press.

Bijker, W.E., Hughes, T.P., Pinch, T. (eds.) (1987) The social construction of technological systems. MIT Press.

Bijker, W.E., Law, J. (eds.) (1992) Shaping technology/building society: Studies in sociotechnical change. MIT Press.

Bolukbasi, T., Chang, K-W., Zou, J.Y., Saligrama, V., Kalai, A.T. (2016) Man Is to Computer Programmer as Woman Is to Homemaker? Debiasing Word Embeddings, 30th Conference on Neural Information Processing Systems (NIPS 2016), Barcelona, Spain.

Bonneau, V., Probst, L., Lefebvre, V. (2018) The Rise of Virtual Personal Assistants. Digital Transformation Monitor, European Commission.

Bowker, G.C. (2000) Bio Diversity Data Diversity. Social Studies of Science, 30(5) (October 2000), 643–83.

Bowker, G.C. (2005) Memory Practices in the Sciences. The MIT Press.

Bowker, G.C. (2014) The Theory/Data Thing. International Journal of Communication, 8 (2043), 1795–1799.

Bowker, G.C., Baker, K., Millerand F., Ribes D. (2010) Toward Information Infrastructure Studies: Ways of Knowing in a Networked Environment. In J. Hunsinger, L. Klastrup, M.M. Allen, M. Matthew (eds.), International Handbook of Internet Research. Springer Netherlands.

Bowker, G.C., Star, S.L. (2000) Sorting Things out: Classification and Its Consequences. Inside Technology. Cambridge. MIT Press.

Brey, P. (2000) Disclosive computer ethics. Computer and Society, 30(4), 10–16. https://doi.org/10.1145/572260.572264

Brey, P. (2010) Values in technology and disclosive ethics. In L. Floridi (ed.) The Cambridge Handbook of Information and Computer Ethics (pp. 41–58). Cambridge University Press.

Brighenti, A.M. (2010) New media and networked visibilities. In A.M. Brighenti (ed.), Visibility in social theory and social research (pp. 91–108). Palgrave Macmillan. https://doi.org/10.1057/9780230282056_4

Bødker, C. (1 September 2014) En ny historie om min krop. Friktion. https://friktionmagasin.dk/en-ny-historie-om-min-krop-979a9b1fefc2

Brøgger, K. (2018) The performative power of (non)human agency assemblages of soft governance. International Journal of Qualitative Studies in Education 31(5), 353–366.

Bygrave, L.A., Bing, J. (2009) Internet Governance Infrastructure and Institutions. Oxford University Press.

Brooker, K. (17 September 2019) Google's quantum bet on the future of AI—and what it means for humanity. FastCompany. https://www.fastcompany.com/90396213/google-quantum-supremacy-future-ai-humanity

Brousseau, E., Marzouki, M. (2012) Internet governance: Old issues, new framings, uncertain implications. In E. Brousseau, M. Marzouki, and C. Méadel (eds.), Governance, Regulation and Powers on the Internet (pp. 368–397). Cambridge University Press. https://doi.org/10.1017/CBO9781139004145.023

Browne, S. (2015) Dark Matters: On the Surveillance of Blackness. Duke University Press.

Bruhn Jensen, K. (2021) A Theory of Communication and Justice. Routledge.

Bryson, J.J. (2018) Patience is not a virtue: the design of intelligent systems and systems of ethics. Ethics and Information Technology 20, 15–26 (2018). https://doi.org/10.1007/s10676-018-9448-6

Bynum, T. (2010) The historical roots of information and computer ethics. In F. Floridi (ed.) Information and Computer Ethics. Cambridge University Press.

Cadwalladr, C. (7 May 2017) The Great British Brexit robbery: how our democracy was hijacked. The Guardian. https://www.theguardian.com/technology/2017/may/07/the-great-british-brexit-robbery-hijacked-democracy

Callon, M., Latour, B. (1992) Don't Throw the Baby Out with the Bath School! A Reply to Collins and Yearley. In A. Pickering (ed.) Science as Practice and Culture. Chicago University Press, 343–368.

Castells, M. (2010) The Rise of the Network Society (Second edition). Wiley Blackwell.

Cavoukian, A. (2009) Privacy by design. The 7 foundational principles. Information and Privacy Commissioner.

Cellan-Jones, R. (2 December 2014) Stephen Hawking warns artificial intelligence could end mankind. BBC. https://www.bbc.com/news/technology-30290540

CEPEJ (2018) European Ethical Charter on the Use of Artificial Intelligence in Judicial Systems and their environment. European Commission for the Efficiency of Justice. https://rm.coe.int/ethical-charter-en-for-publication-4-december-2018/16808f699c

Chen, A. (23 October 2014) The Laborers Who Keep Dick Pics and Beheadings Out of Your Facebook Feed. Wired. https://www.wired.com/2014/10/content-moderation/

Christl, W., Spiekerman, S. (2016) Networks of Control A Report on Corporate Surveillance, Digital Tracking, Big Data and Privacy. Facultas Verlags und Buchhandels AGfacultas Universitätsverlag

Chung, H., Iorga, M., Voas, J., Lee, S. (2017) 'Alexa, Can I Trust You?' Computer, 50(9), 100–104. 10.1109/MC.2017.3571053

Ciccarelli, R. (2021), *Labour Power Virtual and Actual in Digital Production*, Translation by Emma Catherine Gainsforth, Springer Nature.

Clarke, R. (2018) Information Technology and Dataveillance. In T. Monahan, D. M. Wood, Surveillance Studies: A Reader. Oxford University Press, 243–248.

Coeckelbergh, M. (2020) AI Ethics. The MIT Press.

Cohen, J.E. (2012) Configuring the Networked Self: Law, Code, and the Play of Everyday Practice. Yale University Press

Cohen, J.E. (2013) What privacy is for. Harvard Law Review, 126(7).

Collins, H.M. (1987) Expert Systems and the Science of Knowledge. In W.E. Bijker, T.P. Hughes, T. Pinch (eds.) The social construction of technological systems (pp. 329–348). MIT Press.

Collins, H.M., Yearley, S. (1992) Epistemological Chicken. In Andrew Pickering (ed.) Science as Practice and Culture. University of Chicago Press.

Craglia M. (ed.), de Nigris S., Gómez-González, E., Gómez E., Martens B., Iglesias M., Vespe M., Schade S., Micheli M., Kotsev A., Mitton I., Vesnic-Alujevic L., Pignatelli F., Hradec J., Nativi S., Sanchez I., Hamon R., Junklewitz H. (2020) Artificial Intelligence and Digital Transformation: Early Lessons from the COVID 19 Crisis. European Commission. Joint Research Centre. LU: Publications Office. https://data.europa.eu/doi/10.2760/166278

Crevier, D. (1993) AI: The Tumultuous History of the Search for Artificial Intelligence. Basic Books.

Curle, C. T. (2007) Humanité: John Humphrey's Alternative Account of Human Rights. University of Toronto Press.

Danish Business Authority (March 12 2018) The Danish government appoints new expert group on data ethics [Press release]. https://eng.em.dk/news/2018/marts/the -danish-government-appoints-new-expert-group-on-data-ethics

De Hert, P., Gutwirth, S. (2006) Privacy, data protection and law enforcement. Opacity of the individual and transparency of power, in E. Claes, A. Duff and S. Gutwirth (eds.), Privacy and the criminal law, Intersentia, 61–104.

de Wachter, M.A.M. (1997). The European Convention on Bioethics. Hastings Center Report, 27(1), 13–23. https://onlinelibrary.wiley.com/doi/full/10.1002/j.1552-146X .1997.tb00015

Delcker, J., Smith-Meyer, B. (16 January 2020) EU considers temporary ban on facial recognition in public spaces. Politico. https://www.politico.eu/article/eu-considers -temporary-ban-on-facial-recognition-in-public-spaces/

Delacroix, S., Lawrence, N.D. (2019) Bottom-up Data Trusts: Disturbing the 'One Size Fits All' Approach to Data Governance. International Data Privacy Law, 9(4) (November), 236–252. https://doi.org/10.1093/idpl/ipz014

Deleuze, G. (1986) Conversation with Didier Eribon. Le Nouvel Observateur, 23 August 1986. https://onscenes.weebly.com/art/life-as-a-work-of-art

Deleuze, G. (1991) Bergsonism. Translated by H. Tomlinson, B. Habberjam. Urzone, Zone Books. (Originally published in French, 1966).

Deleuze, G. (1992) Postscript on the societies of control. October, 59, 3–7.

Deleuze, G., Guattari, F. (2004) A Thousand Plateaus: Capitalism and Schizophrenia. Continuum (Originally Published in French, 1980).

DeNardis, L. (2012) Hidden Levers of Internet Control. Information, Communication and Society, 15(5), 720–738. https://doi.org/10.1080/1369118X.2012.659199

D'Ignazio, C., Klein, L.F. (2020) Data Feminism. The MIT Press.

Dietrich, E. (2011) Homo Sapiens 2.0: Building the Better Robots of Our Nature. In M., Anderson, S.L. Anderson (eds.) Machine Ethics (pp. 531–537). Cambridge University Press.

Dignum, V., Lopez-Sanchez, M., Micalizio, R., Pavón, J., Slavkovik, M., Smakman, M., van Steenbergen, M. et al. (2018) Ethics by Design: Necessity or Curse? In Proceedings of the 2018 AAAI/ACM Conference on AI, Ethics, and Society - AIES '18, 60–66. ACM Press. https://doi.org/10.1145/3278721.3278745.

Donath, J.S. (1999) Identity and Deception in the Virtual Community. In P. Kollock, M. Smith (eds) Communities in Cyberspace. Routledge.

Dunn, E.C. (2009) Standards without Infrastructure. In S.L. Star, M. Lampland (eds.) Standards and their Stories. Cornell University Press.

Eadicicco, L. (15 April 2016) Meet the Google Exec Trying to Save the Planet. Time. https://time.com/4295351/rebecca-moore-google-earth-outreach/

Edwards, P. (2002) Infrastructure and modernity: scales of force, time, and social organization in the history of sociotechnical systems. In T.J. Misa, P. Brey, A. Feenberg (eds.) Modernity and Technology (pp. 185–225) MIT Press.

Elish, M.C., boyd, d. (2018) Situating methods in the magic of big data and artificial intelligence. Communication Monographs, 85(1), 57–80. DOI: 10.1080/03637751.2017.1375130

Epstein, D. (2013) The making of institutions of information governance: the case of the Internet Governance Forum. Journal of Information Technology 28(2), 137–149.

Epstein, D., Katzenbach, C. and Musiani, F. (2016) Doing internet governance: practices, controversies, infrastructures, and institutions. Internet Policy Review, 5(3) DOI: 10.14763/2016.3.435

Epstein, S. (2008) Culture and Science/Technology: Rethinking Knowledge, Power, Materiality, and Nature. The ANNALS of the American Academy of Political and Social Science, 619(1) (September 2008), 165–82. https://doi.org/10.1177/0002716208319832

Ess, C.M. (2014) Digital Media Ethics. Polity Press.

Eubanks, V. (2018) Automating Inequality: How High-Tech Tools Profile, Police and Punish the Poor. St. Martin's Press.

Eurobarometer 76 (2011). https://ec.europa.eu/commfrontoffice/publicopinion/archives/eb/eb76/eb76_media_en.pdf

Eurobarometer 92 (2019). https://op.europa.eu/en/publication-detail/-/publication/c2fb9fad-db78-11ea-adf7-01aa75ed71a1/language-en

European Commission: see separate list of Policy Documents below.

European Data Protection Supervisor (EDPS) (2015) Towards a New Digital Ethics Data Dignity and Technology.

European Data Protection Supervisor (EDPS) Ethics Advisory Group (2018) Towards a Digital Ethics.

Financial Stability Board (2017) Artificial intelligence and machine learning in financial services: Market developments and financial stability implications. https://www.fsb.org/wp-content/uploads/P011117.pdf

Fjeld, J., Hilligoss, H., Achten, N., Daniel, M.L., Feldman, J., Kagay, S. (2019) Principled artificial intelligence: A map of ethical and rights-based approaches. https://ai-hr.cyber.harvard.edu/primp-viz.html

Flanagan, M., Howe, D.C., Nissenbaum, H. (2008) Embodying values in technology – theory and practice. In J. van den Hoven, J. Weckert (eds.), Information Technology and Moral Philosophy (pp. 322–353). Cambridge University Press.

Floridi, L. (1999) Philosophy and Computing: An Introduction. Routledge.

Floridi, L. (2013) The Ethics of Information. Oxford University Press.

Floridi, L., Cowls, J., Beltrametti, M., Chatila, R., Chazerand, P., Dignum, V., Luetge, C., Madelin, R., Pagallo, U., Rossi, F., Schafer, B., Valcke, P., and Vayena, E. (2018) AI4People White Paper: Twenty Recommendations for an Ethical Framework for a Good AI Society. Minds and Machines, December 2018.

Flyverbom, M. (2011) The Power of Networks Organizing the Global Politics of the Internet. Edward Elgar.

Flyverbom, M. (2019) The Digital Prism: Transparency and Managed Visibilities in a Datafied World. Cambridge University Press.

Foucault, M. (2018) Discipline and Punish: The Birth of the Prison. Translated by Alan Sheridan. In T. Monahan, D.M. Wood, Surveillance Studies A Reader (pp. 36–42). Oxford University Press. (Originally published in French, 1975).

Franklin, M. (2019) Human Rights Futures for the Internet. In B. Wagner, M. Kettemann, K. Vieth (eds.) Research handbook on human rights and digital technology: global politics, law and international rights. Edward Elgar.

Franklin, R.W. (1998) The Poems of Emily Dickinson. The Belknap Press of Harvard University Press.

Friedman, B. (1996) Value-sensitive design. ACM Interactions, 3(6), 17–23.

Friedman, B., Kahn, P.H., Jr., and Borning, A. (2006) Value sensitive design and information systems. In P. Zhang, D. Galletta (eds.), Human-computer interaction in management information systems: Foundations, M.E. Sharpe (pp. 348–372).

Friedman, B., Hendry, G. (2019) Value Sensitive Design Shaping Technology with Moral Imagination. MIT Press.

Friedman, B., Nissenbaum, H. (1995) Minimizing bias in computer systems. In Conference companion of CHI 1995 conference on human factors in computing systems. ACM Press, 444.

Friedman, B., Nissenbaum, H. (1996) Bias in Computer Systems. ACM Transactions on Information Systems, 14(3), 330–47.

Friedman, B., Nissenbaum, H. (1997) Software agents and user autonomy. In Proceedings of first international conference on autonomous agents, ACM Press (pp. 466–469).

Frischmann, B., Selinger, E. (2018) Re-Engineering Humanity, Cambridge University Press.

Frohmann, B. (2007) Foucault, Deleuze, and the ethics of digital networks. In R. Capurro, J. Frühbauer, T. Hausmanninger (eds.), Localizing the Internet. Ethical Aspects in Intercultural Perspective (pp. 57–68). Fink.

Galic, M., Timan, T., Koops, B-J. (2017) Bentham. Deleuze and Beyond: An Overview of Surveillance Theories from the Panopticon to Participation. Philosophy and Technology 30 (1), pp. 9–37.

Gill, E. (13 August, 2020) 'I am expected to just live with these unfair grades' - Student's open letter to the government as she slams A-level results system. Manchester Evening News. https://www.manchestereveningnews.co.uk/news/greater-manchester-news/a-level-results-unfair-downgraded-18764743

Gill, T.G. (1995) Early Expert Systems: Where Are They Now? MIS Quarterly 19(1) (March 1995), 51. https://doi.org/10.2307/249711

Gilpin, L.H., Bau, D., Yuan, B.Z., Bajwa, A., Specter, M., Kagal, L. (2018) Explaining Explanations: An Overview of Interpretability of Machine Learning. 2018 IEEE 5th International Conference on Data Science and Advanced Analytics (DSAA).

Gilroy, P. (2012) There ain't no black in the Union Jack: the cultural politics of race and nation. Routledge. (Originally published in 1987).

Gilroy, P. (2012) British Cultural Studies and the Pitfalls of Identity. In M.G. Durham, D.M. Kellner (eds.) Media and Cultural Studies Keyworks (Second edition) (pp. 337–347). Wiley Blackwell. (Originally published in 1996).

Gray, D.E. (2013) Doing research in the real world. Sage.

Haggerty, K.D., Ericson, R.V. (2000) The surveillance assemblage. British Journal of Sociology, 51(4), 605–622.

Hall, S. (1980) Encoding/Decoding. In S. Hall, D. Hobson, A. Lowe, P. Willis (eds.) Culture, Media, Language Working Papers in Cultural Studies, 1972–79, Hutchinson 118–27. An edited extract from S. Hall, 'Encoding and Decoding in the Television Discourse', cccs stencilled paper no. 7. (Birmingham: Centre for Contemporary Cultural Studies, 1973).

Hall, S. (1994) Cultural Identity and Diaspora. In P. Williams, L. Chrisman (eds.) Colonial Discourse and Post-Colonial Theory A Reader (pp. 222–237). Routledge. (Originally published in 1990).

Hall, S. (1997) The Work of Representation. In Representation: Cultural Representations and Signifying Practices. Sage Publications.

Harvey, D. (1990) The Condition of Postmodernity: An Enquiry into the Origins of Cultural Change. Basil Blackwell.

Harvey, P., Jensen, C.B., Morita, A. (eds.) (2017) Infrastructures and Social Complexity: A Companion. Routledge.

Harraway, D.J. (2016) The Cyborg Manifesto. In Manifestly Haraway, University of Minnesota Press. (Originally published in 1985).

Hasselbalch, G. (2010) Privacy and Jurisdiction in the Global Network Society. https://mediamocracy.files.wordpress.com/2010/05/privacy-and-jurisdiction-in-the -network-society.pdf

Hasselbalch, G. (2013) The Three Momentous Stages of Online Privacy. www .mediamocracy.org. https://mediamocracy.org/2013/08/01/the-three-momentous -stages-of-online-privacy-part-of-my-introduction-to-the-privacy-as-innovation -session-at-the-internet-governance-forum-bali-2013-with-references/

Hasselbalch, G. (2013, B) Privacy is the latest digital media business model (English translation of op ed in Politiken, August 2013) https://mediamocracy.org/2013/08/ 23/data-ethics-the-new-competitive-advantage/

Hasselbalch, G. (2014) Language, Power and Privacy. www.mediamocracy.org. https:// mediamocracy.org/2014/08/26/language-power-and-privacy-talk-at-the-indie-tech -summit-brighton-july-2014/

Hasselbalch, G. (2015) Society of the Destiny Machine and the Algorithmic God(s), www.mediamocracy.org. https://mediamocracy.org/2015/05/14/society-of-the -destiny-machine-and-the-algorithmic-god-s/

Hasselbalch, G. (2018) Let's Talk about AI. AI Alliance Forum. Reposted on Linkedin. https://www.linkedin.com/pulse/lets-talk-ai-gry-hasselbalch/

Hasselbalch, G. (2019) Making sense of data ethics. The powers behind the data ethics debate in European policymaking. Internet Policy Review, 8(2).

Hasselbalch, G. (2020) Culture by Design: A Data Interest Analysis of the European AI Policy Agenda. First Monday, 25(12) (7 December 2020). https://dx.doi.org/10 .5210/fm.v25i12.10861

Hasselbalch, G. (2021) A framework for a data interest analysis of artificial intelli- gence. First Monday, 26(7) (5 July 2021) doi: http://dx.doi.org/10.5210/fm.v26i7 .11091

Hasselbalch, G., Jørgensen, R.F. (2015) Youth, privacy and online media: Framing the right to privacy in public policy-making. First Monday, 20(3) https://doi.org/10 .5210/fm.v20i3.5568

Hasselbalch, G., Olsen, B.K., Tranberg, P. (2020) White Paper on Data Ethics in Public Procurement. DataEthics.eu. https://dataethics.eu/wp-content/uploads/dataethics -whitepaper-april-2020.pdf

Hasselbalch, G., Tranberg, P. (2016) Data Ethics: The New Competitive Advantage, Publishare.

Hasselbalch, G., Tranberg, P. (27 September 2016) Personal Data Stores Want to Give Individuals Power Over Their Data. DataEthics.eu. https://dataethics.eu/personal -data-stores-will-give-individual-power-their-data/

Hasselbalch, G., Tranberg, P. (26 December 2016) Privacy is still alive and kicking in the digital age, TechCrunch. https://techcrunch.com/2016/12/25/privacy-is-still -alive-and-kicking-in-the-digital-age/

Hasselbalch, G., Tranberg, P. (20 May 2020) Contact Tracing Apps are Not Just a Privacy Tech Issue. It's a Question about Power. DataEthics.eu. https://dataethics .eu/contact-tracing-apps-are-not-just-a-privacy-tech-issue-its-a-question-of-power/

Havens, J.C. (2016) *Heartificial Intelligence - Embracing Our Humanity to Maximize Machines,* Penguin Random House.

Hayes, B. (2012) The Surveillance-Industrial Complex. In K. Ball, K.D. Haggerty, D. Lyon (eds.) Routledge Handbook of Surveillance Studies. Routledge.

Heremobility (2020) Barcelona Smart City: By the People, for the People. https://mobility.here.com/learn/smart-city-initiatives/barcelona-smart-city-people-people

Hern, A. (14 August 2020) Do the maths: why England's A-level grading system is unfair. The Guardian. https://www.theguardian.com/education/2020/aug/14/do-the-maths-why-englands-a-level-grading-system-is-unfair

Hern, A. (21 August 2020) Ofqual's A-level algorithm: why did it fail to make the grade? The Guardian. https://www.theguardian.com/education/2020/aug/21/ofqual-exams-algorithm-why-did-it-fail-make-grade-a-levels

Hildebrant, M. (2016) Smart Technologies and the End(s) of Law. Novel Entanglements between Law and Technology. Edward Elgar.

Hill, K. (3 August 2020) Wrongfully Accused by an Algorithm. New York Times. https://www.nytimes.com/2020/06/24/technology/facial-recognition-arrest.html

HLEG: for all European Commission High-Level Expert Group documents, see separate list below.

Hof, S. van der, Lievens, E., Milkaite, I. (2019) The protection of children's personal data in a data-driven world. A closer look at the GDPR from a children's rights perspective. In T. Liefaard, S. Rap, P. Rodrigues (eds.) Monitoring Children's Rights in the Netherlands. 30 Years of the UN Convention on the Rights of the Child. Leiden University Press.

Hoffmann, J., Katzenbach, C., Gollatz, K. (2017) Between coordination and regulation: finding the governance in Internet governance. New Media and Society, 19(9), 1406–1423.

Holten, E. (1 September 2014) SAMTYKKE, Friktion. https://friktionmagasin.dk/samtykke-14841780be52

Hu, Y., Li, W., Wright, D., Aydin, O., Wilson, D., Maher, O., and Raad, M. (2019) Artificial Intelligence Approaches. In J.P. Wilson (ed.) The Geographic Information Science and Technology Body of Knowledge (3rd Quarter 2019 Edition). https://doi.org/10.22224/gistbok/2019.3.4

Hughes, T.P. (1983) Networks of power: Electrification in Western society 1880–1930. The John Hopkins University Press.

Hughes, T.P. (1987) The evolution of large technological systems. In W.E. Bijker, T.P. Hughes, T.Pinch (eds.) The social construction of technological systems (pp. 51–82). MIT Press.

in 't, Veld, S. (26 January 2017) European Privacy Platform [video file]. https://www.youtube.com/watch?v=8_5cdvGMM-U

Jameson, F. (1991) Postmodernism, or, The cultural logic of late capitalism. Duke University Press.

Jankelevitch, V. (2005) Forgiveness. Translated by A. Kelley. University of Chicago Press. (Originally published in 1967).

Jobin, A., Lenca, M., and Vayena, E. (2019) The global landscape of AI ethics guidelines. Nat Mach Intell, 1, 389–399. https://doi.org/10.1038/s42256-019-0088-2

Johnson, B. (11 January 2010) Privacy is no longer a social norm, says Facebook founder, The Guardian. https://www.theguardian.com/technology/2010/jan/11/facebook-privacy

Jørgensen, R.F., Hasselbalch, G., Leth, V. (2013) FOKUSGRUPPE-UNDERSØGELSEN: UNGES PRIVATE OG OFFENTLIGE LIV PÅ SOCIALE MEDIER, Tænketanken Digitale Unge https://www.medieraadet.dk/files/docs/2018–03/Rapport_Unges-private-og-offentlige-liv-paa-sociale-medier_november-2013.pdf

Jørgensen, R.F. (2019) Introduction. In R.F. Jørgensen (ed.) Human Rights in the Age of Platforms. MIT Press.

Kern, S. (1983) The Culture of Time and Space 1880–1918. Harvard University Press.

Keymolen, E., Van der Hof, S. (2019) Can I still trust you, my dear doll? A philosophical and legal exploration of smart toys and trust. Journal of Cyber Policy, 4(2), 143–159.

Kind, C. (23 August 2020) The term 'ethical AI' is finally starting to mean something. VentureBeat. https://venturebeat.com/2020/08/23/the-term-ethical-ai-is-finally -starting-to-mean-something/

Kitchin, R., Lauriault, T. (2014) Towards Critical Data Studies: Charting and Unpacking Data Assemblages and Their Work. Social Science Research Network.

Kowalski, R.M., Limper, S.P., Agatston, P.W. (2008) Cyber Bullying. John Wiley and Sons.

Kramer, A.D.I., Guillory, J.E., Hancock, J.T. (2014) Experimental evidence of massive-scale emotional contagion through social networks. PNAS, 111(29) (July 2014).

Krishna, R.J. (2 July 2014) Sandberg: Facebook Study Was 'Poorly Communicated', Wall Street Journal. https://www.wsj.com/articles/BL-DGB-36278

Krzysztof, J., Gao, S., McKenzie, G., Hu, Y, Bhaduri, B. (2020) GeoAI: Spatially Explicit Artificial Intelligence Techniques for Geographic Knowledge Discovery and Beyond. International Journal of Geographical Information Science 34(4) (2 April 2020), 625–36. https://doi.org/10.1080/13658816.2019.1684500.

Kuhn, T. (1970) The Structure of Scientific Revolutions (Second edition). University Chicago Press.

Lakoff, G., Johnsson, M. (1980) Metaphors We Live By. The University of Chicago Press.

Lapenta, F. (2011) Geomedia: on location-based media, the changing status of collective image production and the emergence of social navigation systems. Visual Studies, 26(1), 14–24. https://doi.org/10.1080/1472586X.2011.548485

Lapenta, F. (2017) Using technology-oriented scenario analysis for innovation research in Research Methods in Service Innovation. In F. Sørensen, F. Lapenta (eds.) Research Methods in Service Innovation. Edgar Allen.

Lapenta, F. (2021) Science Technology and Data Diplomacy for Our Common AI Future. A Geopolitical Analysis and Road Map for AI Driven Sustainable Development. In Finance, Education, Work, Healthcare, for Peace and the Planet.

Larkin, B. (2013) The Politics and Poetics of Infrastructure. The Annual Review of Anthropology 42, 327–43.

Larson, J., Mattu, S., Kirchner, L., Angwin, J. (2016) How We Analyzed the COMPAS Recidivism Algorithm. ProPublica. https://www.propublica.org/article/how-we -analyzed-the-compas-recidivism-algorithm

Latonero, M. (2018) Governing Artificial Intelligence: upholding human rights and dignity. Data and Society.

Latour, B. (1992) Where are the missing masses? The sociology of a few mundane artifacts. In W.E. Bijker and J. Law (eds.) Shaping technology/building society: Studies in sociotechnical change (pp. 225–258). MIT Press.

Latour, B., Venn, C. (2002) Morality and Technology. Theory, Culture and Society, 19(5–6) (1 December 2002), 247–60. https://doi.org/10.1177/026327602761899246

Lecher, C. (25 April 2019) How Amazon automatically tracks and fires warehouse workers for 'productivity'. The Verge. https://www.theverge.com/2019/4/25/ 18516004/amazon-warehouse-fulfillment-centers-productivity-firing-terminations

Lefebvre, A. (2013) Human Rights as a Way of Life: On Bergson's Political Philosophy. Stanford University Press.

Lefebvre, H. (1992) The Production of Space. English translation by D. Nicholson-Smith. Basil Blackwell Ltd. (Originally published in French, 1974).

Lehr, D., Ohm, P. (2017) Playing with the Data: What Legal Scholars Should Learn About Machine Learning, UCDL Review 51, 653–717.

Lessig, L. (2006) Code version 2.0. Basic Books.

Levin, S. (8 September 2016) A beauty contest was judged by AI and the robots didn't like dark skin. The Guardian. https://www.theguardian.com/technology/2016/sep/08/artificial-intelligence-beauty-contest-doesnt-like-black-people

Levin, S. (29 March, 2019) 'Bias deep inside the code': the problem with AI 'ethics' in Silicon Valley. The Guardian. https://www.theguardian.com/technology/2019/mar/28/big-tech-ai-ethics-boards-prejudice

Lieber, R. (11 April 2014) Financial Advice for People Who Aren't Rich. The New York Times. https://www.nytimes.com/2014/04/12/your-money/start-ups-offer-financial-advice-to-people-who-arent-rich.html

Lin, T.C.W. (2014) The New Financial Industry. Alabama Law Review 65, 567, Temple University Legal Studies Research Paper No. 2014-11.

Lohr, S. (1 February 2013) The Origins of Big Data: An Etymological Detective Story. New York Times. https://bits.blogs.nytimes.com/2013/02/01/the-origins-of-big-data-an-etymological-detectivestory/?mtrref=www.google.com&gwh=DC6348FBE0A56CB5C7D9B1A6A287C0E1&gwt=pay&assetType=REGIWALL

Lunau, K. (14 October 2013) Google's Ray Kurzweil on the quest to live forever. Maclean's. https://www.macleans.ca/society/life/how-nanobots-will-help-the-immune-system-and-why-well-be-much-smarter-thanks-to-machines-2/

Lynsky, D. (9 October 2019) 'Alexa, are you invading my privacy?' – the dark side of our voice assistants. The Guardian.

Lyon, D. (1994) The Electronic Eye: The Rise of Surveillance Society. University of Minnesota Press.

Lyon, D. (2001) Surveillance Society Monitoring Everyday Life. Open University Press.

Lyon, D. (2007) Surveillance Studies: An Overview. Polity Press.

Lyon, D. (2010) Liquid surveillance: The contribution of Zygmunt Bauman to surveillance studies. International Political Sociology, 4(4), 325–338. https://doi.org/10.1111/j.1749-5687.2010.00109

Lyon, D. (2014) Surveillance after Snowden. Polity Press.

Lyon, D. (2014) Surveillance, Snowden, and Big Data: Capacities, consequences, critique. Big Data and Society. July–December 2014, 1–13.

Lyon, D. (2018) The Culture of Surveillance. Polity Press.

Maedche, A., Legner, C., Benlian, A. et al. (2019) AI-Based Digital Assistants. Bus Inf Syst Eng 61, 535–544.

Mai, J-E. (2019) Situating Personal Information: Privacy in the Algorithmic Age. In R.F. Jørgensen (ed.) Human Rights in the Age of Platforms (pp. 95–116) MIT Press.

Marcu, B-I. (29 April 2021) Eurodac: Biometrics, Facial Recognition, and the Fundamental Rights of Minors. European Law Blog. https://europeanlawblog.eu/2021/04/29/eurodac-biometrics-facial-recognition-and-the-fundamental-rights-of-minors/

Marr, B. (25 July 2017) 28 Best Quotes about Artificial Intelligence. Forbes. https://www.forbes.com/sites/bernardmarr/2017/07/25/28-best-quotes-about-artificial-intelligence/?sh=32fe61454a6f

Martens, B. (2020) Some economic aspects of access to private data for use in the COVID-19 crisis. In Craglia M. (ed.) Artificial Intelligence and Digital

Transformation: Early Lessons from the COVID-19 Crisis (pp. 16–17). European Commission, Joint Research Centre, Publications Office. https://data.europa.eu/doi/10.2760/166278.

Martin. G. (2014) The Second World War: A Complete History. Rosetta Books. Kindle Edition. (Originally published in 1994).

May, T.C. (1992) The Crypto Anarchist Manifesto. https://www.activism.net/cypherpunk/crypto-anarchy.html

Mayer-Schönberger, V., Cukier, K. (2013) Big data: A revolution that will transform how we live, work and think. John Murray.

Mashey, J.R. (1999) Big Data and the Next Wave of InfraStress Problems, Solutions, Opportunities. 1999 Usenix Annual Technical Conference, June 6–11, Monterey, CA. https://static.usenix.org/event/usenix99/invited_talks/mashey.pdf

Mattioli, G. (26 February 2019) What caused the Genoa bridge collapse – and the end of an Italian national myth? The Guardian. https://www.theguardian.com/cities/2019/feb/26/what-caused-the-genoa-morandi-bridge-collapse-and-the-end-of-an-italian-national-myth

McLeod, J. (14 August 2020) Quietly waiting in the background of the pandemic, AI is about to become a big part of our lives. Financial Post. https://financialpost.com/technology/quietly-waiting-in-the-background-of-the-pandemic-ai-is-about-to-become-a-big-part-of-our-lives/wcm/7d9e3d34-e890-4725-a4a8-6ac9b5920dc4/)

McRobbie, A. (2000) Feminism and youth culture. Second edition. Macmillan.

Mehrabi, N., Morstatter, F., Saxena, N., Lerman, K., Galstyan, A. (2019) A Survey on Bias and Fairness in Machine Learning. ArXiv:1908.09635 [Cs], 17 September 2019. http://arxiv.org/abs/1908.09635.

Menzies, T., Pecheur, C. (2005) Verification and Validation and Artificial Intelligence. Advances in Computers, 65, 153–201.

Merz, F. (2019) Europe and the global AI race. CSS analyses in security policy, no. 247.

Metzinger, T. (8 April 2019) Ethics washing made in Europe. Der Tagesspiegel.

Meyrowitz, J. (1985) No Sense of Place: The Impact of the Electronic Media on Social Behavior. Oxford University Press.

Misa, T.J. (1988) How Machines Make History, and How Historians (And Others) Help Them to Do So. Science, Technology, and Human Values, 13(3/4) (Summer – Autumn, 1988), 308–331.

Misa, T.J. (1992) Theories of Technological Change: Parameters and Purposes. Science, Technology, and Human Values, 17(1) (Winter, 1992), 3–12.

Misa, T.J. (2009) Findings follow framings: navigating the empirical turn. Synthese, 168, 357–375.

Mittelstadt, B.D. (2017) From Individual to Group Privacy in Big Data Analytics. Philosophy and Technology, 30(4) (1 December 2017), 475–94. https://doi.org/10.1007/s13347-017-0253-7

Mittelstadt, B.D., Allo, P., Taddeo, M., Wachter, S., Floridi, L. (2016) The Ethics of Algorithms: Mapping the Debate. Big Data and Society, July–December 2016, 1–21.

Moor, J.H. (1985) What is computer ethics? Metaphilosophy, 16(4), 266–275.

Moor, J. (2006) The Dartmouth College Artificial Intelligence Conference: The Next Fifty Years. AI Magazine, 27(4), 87–91.

Mueller, M.L. (2010) Networks and States: The Global Politics of Internet Governance. Edited by E.J. Wilson. MIT Press.

Mytton, D. (2020) Hiding greenhouse gas emissions in the cloud. Nature Climate Change, 10(701). https://doi.org/10.1038/s41558-020-0837-6

Nelius, J. (4 September 2020) Amazon's Alexa for Landlords Is a Privacy Nightmare Waiting to Happen. Gizmodo. https://gizmodo.com/amazons-alexa-for-landlords-is -a-privacy-nightmare-wait-1844943607

Nemitz, P. (26 January 2017) European Privacy Platform https://www.youtube.com/ watch?v=8_5cdvGMM-U

Nemitz, P. (2018) Constitutional democracy and technology in the age of artificial intelligence. Philosophical Transactions of the Royal Society A, 376(2133). http:// dx.doi.org/10.1098/rsta.2018.0089

Nissenbaum, H. (2010) Privacy in Context: Technology, Policy and the Integrity of Social Life. Stanford University Press.

Noble, S.U. (2018) Algorithms of Oppression: How Search Engines Reinforce Racism. NYU Press.

Noble, S.U. (2018) Critical Surveillance Literacy in Social Media: Interrogating Black Death and Dying Online. In: Close-Up: Black Images Matter. Black Camera: An International Film Journal, 9(2) (Spring 2018), 147–160. doi: 10.2979/ blackcamera.9.2.10

O'Neil, C. (2016) Weapons of math destruction. How Big Data Increases Inequality and Threatens Democracy. Penguin Random House UK.

Orchard, A. (1997) Dictionary of Norse Myth and Legend. Cassell.

Pasquale, F.A. (2013) The Credit Scoring Conundrum. University of Maryland Legal Studies Research Paper No. 2013–45.

Pasquale, F.A. (2015) The black box society – The secret algorithms that control money and information. Harvard University Press.

Pasquale, F.A. (2018) A Rule of Persons, Not Machines: The Limits of Legal Automation. University of Maryland Legal Studies Research Paper No. 20018–08.

Pasquale, F. (2020) New Laws of Robotics: Defending Human Expertise in the Age of AI. Harvard University Press.

Pickering, A. (ed.) (1992) Science as Practice and Culture. The University of Chicago Press.

Poikola, A., Kuikkaniemi, K., and Honko, H. (2018) Mydata – A Nordic Model for humancentered personal data management and processing. Open Knowledge Finland. https://www.lvm.fi/documents/20181/859937/MyData-nordicmodel/ 2e9b4eb0-68d7-463b-9460-821493449a63?version=1.0

Powles, J. (2015–2018) Julia Powles [Profile]. The Guardian. https://www.theguardian .com/profile/julia-powles

Puschmann, C., Burgess, J. (2014) Metaphors of Big Data. International Journal of Communication 8 (2014), 1690–1709.

Rainey, S., Goujon, P. (2011) Toward a Normative Ethical Governance of Technology. Contextual Pragmatism and Ethical Governance. In R. von Schomberg (ed.) Towards Responsible Research and Innovation in the Information and Communication Technologies and Security Technologies Fields (pp. 48–70). European Commission.

Ratner, G., Gad, C. (2019) Data warehousing organization: Infrastructural experimentation with educational governance. Organization, 26(4), 537–552.

Reidenberg, J. R. (1997) Lex Informatica: The formulation of information policy rules through technology. Texas Law Review, 43, 553–593

Reeves, M. (2017) Infrastructural Hope: Anticipating 'Independent Roads' and Territorial Integrity in Southern Kyrgyzstan. Ethnos, 82(4), 711–737.

Rheault, L., Rayment, E. and Musulan, A. (2019) Politicians in the line of fire: Incivility and the treatment of women on social media. Research and Politics (January–March 2019), 1 –7. https://doi-org.ep.fjernadgang.kb.dk/10.1177/2053168018816228

Richards, N.M., King, J.H. (2014) Big Data Ethics. 49 Wake Forest Law Review, 49, 39.

Rønn, K.V., Søe, S.O. (2019) Is social media intelligence private? Privacy in public and the nature of social media intelligence. Intelligence and National Security, 34(3), 362–378. DOI: 10.1080/02684527.2019.1553701

Rorty, R. (1999) Ethics without Principles. In R. Rorty, Philosophy and Social Hope (pp. 72–92). Penguin Books.

Rosenberg, M., Confessore, N., Cadwalladr, C. (17 March 2018) How Trump Consultants Exploited the Facebook Data of Millions. New York Times. https://www.nytimes.com/2018/03/17/us/politics/cambridge-analytica-trump-campaign.html

Schneier, B. (6 March 2006) The Future of Privacy. Schneier on Security. https://www.schneier.com/blog/archives/2006/03/the_future_of_p.html

Schultz, M. (March 3, 2016) Technological totalitarianism, politics and democracy. https://www.youtube.com/watch?v=We5DylG4szM

Scoles, S. (31 July 2019) It's Sentient. The Verge. https://www.theverge.com/2019/7/31/20746926/sentient-national-reconnaissance-office-spy-satellites-artificial-intelligence-ai

Searle, J.R. (1980) Minds, brains, and programs. Behavioral and Brain Sciences, 3(3), 417–457.

Searle, J.R. (1997) The Mystery of Consciousness. The New York Review of Books.

Seville, H., Field, D.G. (2011) What Can AI Do for Ethics? In M. Anderson, S.L. Anderson (eds.) Machine Ethics (pp. 499–511). Cambridge University Press.

Shapiro, S.P. (2005) Agency theory. Annual Review of Sociology, 31(1), 263–284.

Shilton, K. (2015) Anticipatory Ethics for a Future Internet: Analyzing Values During the Design of an Internet Infrastructure. Science and Engineering Ethics, 21(1) (February 2015), 1–18. https://doi.org/10.1007/s11948-013-9510-z.

Simonite, T. (26 October 2020) How an Algorithm Blocked Kidney Transplants to Black Patients. Wired. https://www.wired.com/story/how-algorithm-blocked-kidney-transplants-black-patients/

Smith, B.C. (2019) The Promise of Artificial Intelligence Reckoning and Judgment. MIT Press.

Smuha, N.A. (2019) The EU approach to ethics guidelines for trustworthy artificial intelligence. Computer Law Review International, 20(4), pp. 97–106.

Smuha, N.A. (2020) Beyond a Human Rights-based approach to AI Governance: Promise, Pitfalls, Plea. http://dx.doi.org/10.2139/ssrn.3543112

Solove, D. (2001) Privacy and Power: Computer Databases and Metaphors for Information Privacy. Stanford Law Review, 53(1393).

Solove, D. (2002) Conceptualizing Privacy. California Law Review, 90(4), 1087–1155. doi:10.2307/3481326

Solove, D. (2008) Understanding Privacy. Harvard University Press.

Solove, D.J. (2006) A Taxonomy of Privacy. University of Pennsylvania Law Review, 154(3), 477. GWU Law School Public Law Research Paper No. 129.

Spiekermann, S., Hampson P., Ess, C. M., Hoff, J., Coeckelbergh, M., Franckis, G. (2017) The Ghost of Transhumanism and the Sentience of Existence. Retrieved from The Privacy Surgeon. http://privacysurgeon.org/blog/wp-content/uploads/2017/07/Human-manifesto_26_short-1.pdf

Spielkamp, M. (ed.) (2019) Automating Society Taking Stock of Automated Decision-Making in the EU. AlgorithmWatch. https://algorithmwatch.org/wp-content/uploads/2019/02/Automating_Society_Report_2019.pdf

Spillman, L., Strand, M. (2013) Interest-oriented action. Annual Review of Sociology, 39(1), 85–104.

Star, L.S. (1999) The Ethography of Infrastructure, The American Behavioral Scientist, Nov/Dec 1999; 43(3), 377–392.

Star, S.L., Bowker, G. C. (2006) How to Infrastructure? In L.A. Lievrouw and S. Livingstone (eds.) Handbook of New Media. Social Shaping and Social Consequences of ICTs, pp. 230–245. Updated student edition. SAGE Publications Ltd.

Stoddart, E. (2012) A surveillance of care: Evaluating surveillance ethically. In K. Ball, K. Haggerty, D. Lyon (eds.) Routledge Handbook of Surveillance Studies (pp. 369–376). Routledge.

Strubell, E., Ganesh, A., McCallum, A. (2019) Energy and Policy Considerations for Deep Learning in NLP, arXiv:1906.02243

Stupp, C. (6 April 2018) Cambridge Analytica harvested 2.7 million Facebook users' data in the EU. Euractiv. https://www.euractiv.com/section/data-protection/news/cambridge-analytica-harvested-2-7-million-facebook-users-data-in-the-eu/

Surur (22 April 2018) Microsoft's AI used to identify potential school drop outs. MsPoweruser. https://mspoweruser.com/microsofts-ai-used-to-identify-potential-school-drop-outs/

Sweeney, L. (2013) Discrimination in Online Ad Delivery, acm queue, 11(3).

The Guardian (1 November 2013) NSA Prism program slides. https://www.theguardian.com/world/interactive/2013/nov/01/prism-slides-nsa-document

The Guardian (2018) The Cambridge Analytica Files. https://www.theguardian.com/news/series/cambridge-analytica-files

The Shift Project (2019) Lean ICT – Towards Digital Sobriety. https://theshiftproject.org/wp-content/uploads/2019/03/Lean-ICT-Report_The-Shift-Project_2019.pdf

Thompson, E. (1979) The making of the English working class. Penguin Books. (Originally published 1963).

Thorpe, B. (1907) (trans.) The Elder Edda of Saemund Sigfusson, and the Younger Edda of Snorre Sturleson. Norroena Society.

Tigard, D.W. (2020) There Is No Techno-Responsibility Gap. Philosophy and Technology, 9 July 2020. https://doi.org/10.1007/s13347-020-00414-7

Tisne, M. (2020) The Data Delusion: Protecting Individual Data Isn't Enough When the Harm is Collective. Stanford Policy Center.

Topham, G. (25 September 2015) Volkswagen scandal – seven days that rocked the German carmaker. The Guardian. https://www.theguardian.com/business/2015/sep/25/vw-emissions-scandal-seven-days

Tranberg, P., Heuer, S. (2013) Fake It: Your Guide to Digital Self Defence. People's Press.

Turing, A. (2004) Computing Machinery and Intelligence. In J.B. Copeland (ed.) The Essential Turing: The Ideas that Gave Birth to the Computer Age. Clarendon Press.

Turkle, S. (1997) Life on the Screen Identity in the Age of the Internet. Touchstone.

UK Government (2018) Digital Charter. https://www.gov.uk/government/publications/digital-charter/digital-charter

Umbrello, S. (2019) Beneficial Artificial Intelligence Coordination by Means of a Value Sensitive Design Approach. Big Data and Cognitive Computing, 3(1), 5. MDPI AG.

Umbrello, S. (2020) Mapping Value Sensitive Design onto AI for Social Good Principles. Preprint.

Umbrello, S., De Bellis, A.F. (2018) A Value-Sensitive Design Approach to Intelligent Agents. In R.V. Yampolskiy (ed.) Artificial Intelligence Safety and Security (pp. 395–410). CRC Press: Boca Raton.

Umbrello, S., Yampolskiy, R.V. (2020) Designing AI for Explainability and Verifiability: A Value Sensitive Design Approach to Avoid Artificial Stupidity in Autonomous Vehicles. Preprint.

Vallor, S. (2016) Technology and the Virtues. Oxford University Press.

Valtysson, B. (2017) Regulating the Void: Online Participatory Cultures, User-Generated Content, and the Digital Agenda for Europe. In P. Meil, V. Kirov (eds.) Policy Implications of Virtual Work (pp. 83–107). Springer International Publishing. https://doi.org/10.1007/978-3-319-52057-5_4.

van Wynsberghe, A. (2021) Sustainable AI: AI for sustainability and the sustainability of AI. AI Ethics. https://doi.org/10.1007/s43681-021-00043-6

van Wynsberghe, A., Robbins, S. (2019) Critiquing the Reasons for Making Artificial Moral Agents. Science and Engineering Ethics (2019) 25, 719–735. https://doi.org/10.1007/s11948-018-0030-8

Vasse'i, R.M. (2019) The Ethical Guidelines for Trustworthy AI – A Procrastination of Effective Law Enforcement. CRi5/2019.

Veale, M. (2019) A Critical Take on the Policy Recommendations of the EU High-Level Expert Group on Artificial Intelligence. European Journal of Risk Regulation. Preprint. Faculty of Laws University College London Law Research Paper, No. 8, 2019.

Veliz, C. (2020) Privacy is Power: Why and How You Should Take Back Control of Your Data. Bantam Press.

Vesnic-Alujevic, L., Pignatelli F. (2020) Privacy, democracy and the public sphere in the age of COVID-19. In M. Craglia (ed.) Artificial Intelligence and Digital Transformation: Early Lessons from the COVID 19 Crisis (pp. 24–26). European Commission, Joint Research Centre, Publications Office. https://data.europa.eu/doi/10.2760/166278

Vestager, M. (9 September 2016) Making Data Work for Us. https://ec.europa.eu/commission/commissioners/2014-2019/vestager/announcements/making-data-work-us_en Video available at https://vimeo.com/183481796

Vinge, V. (2001) True Names: And the Opening of the Cyberspace Frontier. In J. Frenkel (ed.) A Tor Book. Tom Doherty Associates. (Originally Published in 1981).

von der Leyen, U. (2019) A Union that strives for more. My agenda for Europe: Political guidelines for the next European Commission 2019–2024. https://ec.europa.eu/commission/sites/beta-political/files/political-guidelines-next-commission_en.pdf

Wachter, S. (2019) Data Protection in the Age of Big Data. Nature Electronics, 2(1) (1 January), 6–7. https://doi.org/10.1038/s41928-018-0193-y

Wachter, S., Mittelstadt, B., Floridi, L. (2017) Why a Right to Explanation of Automated Decision-Making Does Not Exist in the General Data Protection Regulation. International Data Privacy Law 7(2), 76–99.

Wagner, B. (2018). Ethics as an escape from regulation: from ethics-washing to ethics-shopping? In M. Hildebrandt (Ed.), *Being Profiling. Cogitas Ergo Sum.* Amsterdam: Amsterdam University Press. Retrieved from https://www.privacylab.at/wp-content/uploads/2018/07/Ben_Wagner_Ethics-as-an-Escape-from-Regulation_2018_BW9.pdf

Wagner, B., Kettemann, M., and Vieth, K. (2019) Introduction. In B. Wagner et al. (eds.) Research handbook on human rights and digital technology: global politics, law and international rights. Edward Elgar.

Wahl, T. (10 September 2019) EU Creates New Central Database for Convicted Third Country Nationals. Eucrim. https://eucrim.eu/news/eu-creates-new-central-database-convicted-third-country-nationals/

Warman, M. (8 February 2012) EU Privacy regulations subject to 'unprecedented lobbying'. The Telegraph. https://www.telegraph.co.uk/technology/news/9070019/EU-Privacy-regulations-subject-to-unprecedented-lobbying.html

Watson, S.M. (n.d.) 'Data is the new ...' Dis Magazine. http://dismagazine.com/blog/73298/sara-m-watson-metaphors-of-big-data/

Webster, F. (2014) Theories of the Information Society. 4th edition. Routledge.

WHO (2018) Big data and artificial intelligence for achieving universal health coverage: an international consultation on ethics. https://apps.who.int/iris/bitstream/handle/10665/275417/WHO-HMM-IER-REK-2018.2-eng.pdf?ua=1

Wiener, N. (2013) Cybernetics or, Control and Communication in the Animal and the Machine. Second edition. Martino Publishing. (Originally published 1948).

Wikipedia (2020) List of data breaches. Wikipedia. https://en.wikipedia.org/wiki/List_of_data_breaches

Williams, R. (1993) Culture is ordinary. In A. Gray and J. McGuigan (eds.) Studying culture an introductory reader. Edward Arnold. (Originally published 1958).

Williams, R. (24 June 2020) I was wrongfully arrested because of facial recognition. Why are police allowed to use it? The Washington Post. https://www.washingtonpost.com/opinions/2020/06/24/i-was-wrongfully-arrested-because-facial-recognition-why-are-police-allowed-use-this-technology/

Winfield, A.F.T., Jirotka, M. (2018) Ethical governance is essential to building trust in robotics and artificial intelligence systems. Philosophical Transactions of the Royal Society A: Mathematical, Physical and Engineering Sciences, 376(2133).

Winfield, A.F.T. (28 June 2019) Energy and Exploitation: AIs dirty secrets. Alan Winfield's weblog.

Winner, L. (1980) Do artifacts have politics? Daedalus, 109(1), 121–136.

Winner, L. (1986) The Whale and the Reactor, A Search for Limits in an Age of High Technology. Second Edition. The University of Chicago Press.

Woolgar, S. (1987) Reconstructing Man and Machine: A Note on Sociological Critiques of Cognitivism. In W. E., Bijker, T. P. Hughes, T. Pinch (eds.) The social construction of technological systems (pp. 311–328). MIT Press.

Xu, J., et al. (2021) Stigma, Discrimination, and Hate Crimes in Chinese-Speaking World Amid Covid-19 Pandemic. Asian Journal of Criminology, 16, 51–74. https://doi-org.ep.fjernadgang.kb.dk/10.1007/s11417-020-09339-8

Zanzotto, F.M. (2019) Viewpoint: Human-in-the-loop Artificial Intelligence. Journal of Artificial Intelligence Research 64, 243–252.

Zarsky, T. (2017) Incompatible: The GDPR in the Age of Big Data. Seton Hall Law Review, 47(4/2), 2017.

Zuboff, S. (5 March 2016) The secrets of surveillance capitalism. Frankfurter Allgemeine. http://www.faz.net/aktuell/feuilleton/debatten/the-digital-debate/shoshana-zuboff-secrets-of-surveillance-capitalism-14103616.html

Zuboff, S. (9 September 2014) A digital declaration. Frankfurter Allgemeine. http://www.faz.net/aktuell/feuilleton/debatten/the-digital-debate/shoshan-zuboff-on-big-data-as-surveillance-capitalism-13152525.html

Zuboff, S. (2019) The Age of Surveillance Capitalism: The Fight for a Human Future at the New Frontier of Power. Profile Books.

High-Level Expert Group on AI Documents

High-Level Expert Group on Artificial Intelligence (HLEG A) (2019) Ethics Guidelines for Trustworthy AI. https://ec.europa.eu/digital-single-market/en/news/ethicsguidelines-trustworthy-ai
High-Level Expert Group on Artificial Intelligence (HLEG B) (2019) Policy and investment recommendations for Trustworthy AI. https://ec.europa.eu/digital-single -market/en/news/policyand-Investment-recommendations-trustworthy-artificial -intelligence
High-Level Expert Group on Artificial Intelligence (HLEG C) (2019) A Definition of AI: Main Capabilities and Disciplines. https://ec.europa.eu/futurium/en/ai-alliance -consultation
High-Level Expert Group on Artificial Intelligence (HLEG D) (2018) Draft Ethics Guidelines for Trustworthy AI. Working document, 18 December 2018. https://ec .europa.eu/digital-single-market/en/news/draft-ethics-guidelines-trustworthy-ai
High-Level Expert Group on Artificial Intelligence (HLEG E) (2018) 'Minutes of the first meeting' (27 June). Register of Commission Expert Groups and Other Similar Entities.

Policy Documents

CEPEJ (2018) European Ethical Charter on the Use of Artificial Intelligence in Judicial Systems and their environment, by the European Commission for the Efficiency of Justice (CEPEJ) of the Council of Europe. Adopted at the 31st plenary meeting of the CEPEJ (Strasbourg, 3–4 December 2018)
Council of Europe (1997) Convention for the Protection of Human Rights and Dignity of the Human Being with regard to the Application of Biology and Medicine: Convention on Human Rights and Biomedicine. European Treaty Series - No. 164, Oviedo, 4.IV.1997. https://www.coe.int/en/web/conventions/full-list/-/conventions/treaty/164
Data Ethics Expert Group (2018) Data for the Benefit of the People Recommendations from the Danish Expert Group on Data Ethics. https://dataetiskraad.dk/sites/default/files/2020-02/Recommendations%20from%20the%20Danish%20Expert%20Group%20on%20Data%20Ethics.pdf
European Commission A (2020) Trans-European Transport Network (TEN-T) https://ec.europa.eu/transport/themes/infrastructure_en
European Commission B (2020) Infrastructure and Investment. https://ec.europa.eu/transport/themes/infrastructure_en
European Commission C (2019) The Connecting Europe Facility Five Years Supporting European   Infrastructure.   https://ec.europa.eu/inea/sites/inea/files/cefpub/cef _implementation_brochure_2019.pdf
European Commission D (2010) A Digital Agenda for Europe. COM(2010)245 final, Brussels, 19.5.2010. https://eur-lex.europa.eu/LexUriServ/LexUriServ.do?uri =COM:2010:0245:FIN:EN:PDF

European Commission E (2015) A Digital Single Market Strategy for Europe. COM(2015) 192 final, Brussels, 6.5.2015. https://eur-lex.europa.eu/legal-content/EN/TXT/?uri=COM%3A2015%3A192%3AFIN

European Commission F (2016) Digitising European Industry: Reaping the full benefits of a Digital Single Market. COM(2016) 180 final, Brussels, 19.4.2016 https://eur-lex.europa.eu/legal-content/EN/TXT/?uri=CELEX:52016DC0180

European Commission G (2016) European Cloud Initiative – Building a competitive data and knowledge economy in Europe. COM(2016) 178 final, Brussels, 19.4.2016. https://ec.europa.eu/digital-single-market/en/news/communication-european-cloud-initiative-building-competitive-data-and-knowledge-economy-europe

European Commission H (2020) A European strategy for data. COM (2020) 66 final, Brussels, 19.2.2020. https://ec.europa.eu/info/sites/info/files/communication-european-strategy-data-19feb2020_en.pdf

European Commission I (2020) On Artificial Intelligence – A European approach to excellence and trust. COM(2020) 65 final, Brussels, 19.2.2020. https://ec.europa.eu/info/sites/info/files/commission-white-paper-artificial-intelligence-feb2020_en.pdf

European Commission J (9 March 2018) Call for a High-Level Expert Group on Artificial Intelligence. https://ec.europa.eu/digital-single-market/en/news/call-high-level-expert-group-artificial-intelligence

European Commission K (2018) Artificial intelligence for Europe (25 April), at https://ec.europa.eu/digital-single-market/en/news/communication-artificial-intelligence-europe

European Commission L (2018) Declaration of cooperation on artificial intelligence. https://ec.europa.eu/digital-single-market/en/artificial-intelligence#Declaration-of-cooperation-on-Artificial-Intelligence

European Commission M (2018) Coordinated plan on artificial intelligence 'made in Europe' (7 December) https://ec.europa.eu/commission/presscorner/detail/ro/memo_18_6690

European Commission N (2019) Building trust in human-centric artificial intelligence (9 April) https://ec.europa.eu/digital-single-market/en/news/communication-building-trust-human-centric-artificial-intelligence

European Commission O (2021) Proposal for a Regulation of the European Parliament and of the Council, laying down harmonised rules on artificial intelligence (Artificial Intelligence Act) and amending certain Union Legislative Acts. https://digital-strategy.ec.europa.eu/en/library/proposal-regulation-laying-down-harmonised-rules-artificial-intelligence

European Parliament (16 February 2017) European Parliament resolution of 16 February 2017 with recommendations to the Commission on Civil Law Rules on Robotics. (2015/2103(INL))

European Parliament (12 February 2019) A comprehensive European industrial policy on artificial intelligence and robotics. Strasbourg. https://www.europarl.europa.eu/doceo/document/TA-8-2019-0081_EN.html

OECD (22 May 2019) Principles on Artificial Intelligence. In OECD Council Recommendation on Artificial Intelligence, OECD/LEGAL/0449, 22.05.2019. https://legalinstruments.oecd.org/en/instruments/OECD-LEGAL-0449

Regulation (EU) 2016/679 of the European Parliament and of the Council of 27 April 2016 on the protection of natural persons with regard to the processing of personal data and on the free movement of such data, and repealing Directive 95/46/EC (General Data Protection Regulation) https://rm.coe.int/ethical-charter-en-for-publication-4-december-2018/16808f699c

UN (2013) 68/167 The right to privacy in the digital age. Resolution adopted by the General Assembly on 18 December 2013.
World Summit on the Information Society (2013) Basic Information: About WSIS. https://www.itu.int/net/wsis/basic/about.html

Index